SUPERSTRING THEORY IN THE 21ST CENTURY. HORIZONS IN WORLD PHYSICS. VOLUME 270

HORIZONS IN WORLD PHYSICS

Additional books in this series can be found on Nova's website under the Series tab.

Additional E-books in this series can be found on Nova's website under the E-book tab.

SUPERSTRING THEORY IN THE 21ST CENTURY. HORIZONS IN WORLD PHYSICS. VOLUME 270

GEROLD B. CHARNEY
EDITOR

Nova Science Publishers, Inc.
New York

For permission to use material from this book please contact us:
Telephone 631-231-7269; Fax 631-231-8175
Web Site: http://www.novapublishers.com

LIBRARY OF CONGRESS CATALOGING-IN-PUBLICATION DATA

Superstring theory in the 21st century / editor, Gerold B. Charney.
p. cm. -- (Horizons in world physics ; v. 270)
Includes index.
ISBN 978-1-61668-385-6 (hardcover)
1. Superstring theories. I. Charney, Gerold B. II. Title: Superstring theory in the twenty-first century.
QC794.6.S85S865 2009
539.7'258--dc22

2010001759

Published by Nova Science Publishers, Inc. ✝ New York

CONTENTS

PREFACE

In this book, the authors present and review current data on superstring theory including topics such as: the classification of elementary particles in the phase-trees theory; recent developments in constructing superstring amplitudes for genus g higher than 2; the cosmology of type IIB superstrings theory; thermal and quantum-induced superstring cosmology, and others.

Chapter 1 - Phase-trees theory develops from two principles: 1* Quantum indeterminacy, and 2* phase space expanding. One can reveal classification of elementary particles and classification of interactions in this theory. Fundamental characteristics of them are components of the inflation potential. They are proven and calculated in this theory.

Chapter 2 - The authors will present the recent developments in constructing superstring amplitudes for genus g higher than 2.

The main problem in defining superstring amplitudes is due to the ambiguity of the super covariant path integral formulation. Super Weyl and super diffeomorphism invariance of the theory give rise to certain constraints on the vacuum-to-vacuum amplitudes, which must then behave as suitable modular forms. A strategy for determining such amplitudes is then to impose physical constraints to opportune subspaces of modular forms selected by the above symmetries. Recently, this method has been applied to determine the vacuum-to-vacuum amplitudes for superstrings at genus three and four.

Here the authors will explain the mathematical and physical details underlying such constructions, with particular attention to the $g=3$ case. Differently from the bosonic case, it will be shown that the main ingredients are certain equivariant modular forms supporting finite groups having a physical meaning. The physical constraints can then be recasted in terms of representation theory and the problem reduces itself in selecting the invariants.

The authors will show that the constraints together with some simple assumptions give rise to a unique solution. However, the authors also provide some criticism, evidencing some related open problems.

Chapter 3 - In the typical superstrings cosmology, the dilaton is usually interpreted as a Quintessence field. This work is a review of an alternative interpretation of the dilaton, namely, as the dark matter of the universe, in the context of a particular cosmological model derived from type IIB supergravity theory with fluxes. First the authors study the conditions needed to have an early epoch of inflationary expansion with a potential coming from type IIB superstring theory with fluxes involving two moduli fields; the dilaton and the axion. The phenomenology of this potential is different from the usual hybrid inflation scenario and the authors analyze the possibility that the system of field equations undergo a period of inflation in three different regimes with the dynamics modified by a Randall-Sundrum term in the Friedmann equation. The authors find that the system, can produce inflation and due to the modification of the dynamics, a period of accelerated contraction can follow or preceed this inflationary stage depending on the sign of one of the parameters of the potential. The authors discuss on the viability of this model in a cosmological context. With this alternative interpretation they also find that the model gives a similar evolution and structure formation of the universe compared with the _CDM model in the linear regime of fluctuations of the structure formation. Some free parameters of the theory are fixed using the present cosmological observations. In the nonlinear regime there are some differences between the type IIB supergravity theory with the traditional CDM paradigm. The supergravity theory predicts the formation of galaxies earlier than the CDM and there is no density cusp in the centre of galaxies. These differences can distinguish both models and might give a distinctive feature to the phenomenology of the cosmology coming from superstring theory with fluxes.

Chapter 4 - In this paper the authors show that the holomorphic representation is appropriate for description in a consistent way string and string field theories, when the considered number of component fields of the string field is finite. A new Lagrangian for the closed string is obtained and shown to be equivalent to Nambu-Goto's Lagrangian. The authors give the notion of anti-string, evaluate the propagator for the string field, and calculate the convolution of two of them.

Chapter 5 - In this work, the authors review the results of two of their references dedicated to the description of the early Universe cosmology induced by quantum and thermal effects in superstring theories. The present evolution of the Universe is described very accurately by the standard Λ-CDM scenario, while very little is known about the early cosmological eras. String theory provides a consistent microscopic theory to account for such missing epochs. In our framework, the Universe is a torus filled with a gas of superstrings. The authors first show how to describe the thermodynamical properties of this system, namely energy density and pressure, by introducing temperature and supersymmetry breaking effects at a fundamental level by appropriate boundary conditions.

The authors focus on the intermediate period of the history: After the very early "Hagedorn era" and before the late electroweak phase transition. The authors determine the back-reaction of the gas of strings on the initially static space-time, which then yields the induced cosmology. The consistency of our approach is guaranteed by checking the quasi-staticness of the evolution.

It turns out that for arbitrary initial boundary conditions at the exit of the Hagedorn era, the quasi-static evolutions are universally attracted to radiation-dominated solutions. It is shown that at these attractor points, the temperature, the inverse scale factor of the Universe and the supersymmetry breaking scale evolve proportionally. There are two important effects which result from the underlying string description. First, initially small internal dimensions can be spontaneously decompactified during the attraction to a radiation dominated Universe. Second, the radii of internal dimensions can be stabilized.

Chapter 6 - Nambu-Goto proposed a simple open string as a sequel to the dual resonance model. Afterwards, Lovelace had shown that the physical states have negative norms except in 26 dimension. The discovery of Scherck and Schwartz that the spectrum contains graviton, raised the ultimate hope that all the four interactions, observed in nature can be unified. But there are also the unphysical tachyons which were subsequently found to vanish in ten dimensional superstring constructed by Green and Schwartz. Then came the pheno-menological rise in both conceptual and mathematical complexity and the simplicity of the Nambu-Goto string was forgotten. This paper and few others attempt at a successful revival of the classic idea. First of all, noting the Mandelstam's equivalence of one bosonic mode to two fermionic modes, the Nambu-Goto string in 26 dimension is written as the four coordinates and 4 groups of eleven fermions in SO(3,1) bosonic representations. It is proved that the world sheet action is supersymmetric, rich in particle quanta and is equivalent to a four dimensional superstring. Constructing the supersymmetric charge, the Hamiltonian is found to allow only positive energy states, proving that there are no tachyons. There is also manifest modular invariance.

At first, there appears to be many ghosts and negative norm states. After quantizing in NS and R formulations, the authors construct the Virasoro operators, the superalgebra and the physical states. The BRST supercharge, found by using conformal and superconformal ghosts, is nilpotent. The Theory is ghost free, anomaly free and is unitary.

With a definite proof that the Nambu-Goto string in a superstring formulation is correct, the authors see that the symmetry of action $SO(6)$ $SO(5)$ is also the gauge symmetry and can descend to three generations of Supersymmetric standard model by the use of Wilson loop. For completeness, the authors construct the particle spectrum, consisting of the Higgs sector, Gauge sector and the fermions in detail. As an interesting addition, they show that the particles of the standard model are superpartners of each other, following Nambu and Witten. From a simple Nambu-Goto superstring action, the authors construct the graviton and gravtino states which was first ever written down action of N=1 supergravity.

In: Superstring Theory in the 21st Century…
Editor: G. B. Charney, pp. 1-11
ISBN: 978-1-61668-385-6

Chapter 1

CLASSIFICATION OF ELEMENTARY PARTICLES IN THE PHASE-TREES THEORY

A.V. Khromov
VNII OFI – Institute of Optical and Physical Measurements,
Moscow, Russia

ABSTRACT

Phase-trees theory develops from two principles: 1* Quantum indeterminacy, and 2* phase space expanding. One can reveal classification of elementary particles and classification of interactions in this theory. Fundamental characteristics of them are components of the inflation potential. They are proven and calculated in this theory.

The string theory is the maturity of the Standard Model. But to choose mechanical models, builds, size, string tension, are not in themselves better than the fundamental theory. In contrast with M-theory, the foundation of General Relativity was in geometry and symmetry. In the phase-trees theory, the phase tree is also not a mechanical model. It may truly be called the logical formula or combination of two principles:

1* Quantum indeterminacy, and 2* Phase space expanding. These two principles are trustworthy and they are an all-sufficient base to seek physical consequences.

Classical physics explained (before A. Fridman and E. Hubble) expanding 2* very well as a "law of nature". There exists one charismatic thing — time T. It has intrinsic character or ability, or willpower to grow: $T=T_0+\Delta T$. And it is the whip-cracker of all movements and changes. Not only does it compel change : $S = V*T$; $a = \exp(HT)$ etc., but it determines how to change. It is a very manful theory, but determinism is not compatible with quantum indeterminacy 1*.

Quantum mechanics explains indeterminacy 1* very decidedly as wave function or by path integral formulation. But in quantum mechanics, space does not expand. Can we devise some other theory (some other artificial universe) that rigorously and organically develops from both conditions 1* and 2* and combine both of them? The phase-trees theory and its variants [1–6] are attempt to devise such a theory. It can not begin with some unexplained parameters or laws. Meanwhile quantum indeterminacy 1* has special preference. Quantum indeterminacy as real chaos is not something to explain and vindicate. One can not ask about real chaos: how great is it? or what qualities and hidden parameters does it have? or why is it such and not other? or is it finite or infinite? Real chaos is real singularity, exotic ability to scatter and unflate, singularity of inflation potential. Since relativity and inflation, physical reality is more an ensemble of events then a throng of contraptions.

Phase tree is a recordation (scenario) of all scatterings (1*) and of all dilatations (2*) to generate a physical object or to attain a given phase of cosmological inflation. It is an accidental branchy process. One can diagrammatically project this process by phase trees, then by the cipher codes and then simulate it by computer. Phase tree is a binary tree of special type: in the root and in each node, right part is subtree of the left part. There are finite number of H-storey phase trees P(H). We can sequentially construct and number all of them. Eigensubtrees B_i of A are events. Phase tree A is event and space of events. Sum of all Bi constitute inflation potential Ω of A and they are phase space of A. Phase tree A randomly scatters in one of eigensubtrees B_x and frames new phase tree A^Bx. New phase A^Bx comprise all eigensubtrees of A and Bx, and therefore phase space expands (heating). Moreover at the root of A^Bx arise many new subtrees and Ω grew by divertifications. Frequency $F_A(B)$ of subtree B in the tree A is the number of methods to superimpose B on A. Probability Pr(A^B) to scatter A into eigensubtree B is

$$\Pr(A\wedge B) = F_A(B) / \Omega, \tag{1}$$

in accordance with quantum theory 1*.

Each H-storey phase tree P(R,H) of range R ($R \leq H\text{-}2$) has rigorous constructive characteristics. Range of the tree is R if it has chain of range R and not more then R. Range R of the chain means that this chain turns off the chain of range R-1. Range of the first chain is one. Phase tree has individual spectrum of eigensubtreees and their frequencies. Sum of frequencies of all eigensubtrees is inflation potential Ω of the tree. Range R is specific symmetry of the phase tree, analogous to gauge symmetry. It regulates emergence of new particles and interactions. Inflation potential is $\Omega = \Omega_0 + \Omega_1 + \Omega_2 + \Omega_{(R-1)}\ (+\Omega_R)$. These characteristics are rigorous mathematical objects. They are fundamental constants in this theory. We acquire them not as fenomenons or befitting parameters (externality). They emerge rigorously from 1* and 2* only. They are proven and calculated in this theory. Inflation potential Ω is the measure of the phase space. Part Ω_0 of range 0 is nearly additive and conservative in the phase space (like energy).

There are H-storey phase trees EP(R,H) of special type, with minimal Ω. They do not have subtrees of range R. They can be excited and they interact by exchanging subtrees of range < R, but they cannot create phase of range R+1 before they multiply. They are elementary phases and elementary particles in this theory. One can reveal classification of elementary particles and classification of interactions in this theory. Phases of range 0–3 constitute 3- dimensional space. Phase trees and spaces of range > 3 emerge as detached crumbs in 3-dimensional space. This is compactification of multidimensional spaces. We really see this compactification when we imitate process (1*-2*) by computer.

Most probable to arise are elementary particles EP(R,R+2).

ELEMENTARY PARTICLES EP(4,6) OF RANGE FOUR

{...} —cipher code of the phase tree. Ω0, Ω1, Ω2, Ω3 —parts of the inflation potential Ω.

Pr —probability to come about by four first steps of inflation (direct emergence).

{9,6,27,29,9,62,10,28,8}, Ω_0 =44 ; Ω_1=54; Ω_2=149; Ω_3=142;
Pr = $6.413*10^{-5}$,

{10,6,23,11,29,9,62,10,28,8}, Ω_0 =44; Ω_1=53; Ω_2=179; Ω_3=174;
Pr = $8.354*10^{-5}$,

{10,6,23,11,29,9,62,22,28,8}, Ω_0 =44 ; Ω_1=53; Ω_2=175; Ω_3=134;
Pr = $8.354*10^{-5}$,

{11,6,23,27,61,21,62,22,10,60,20}, Ω_0 =52; Ω_1=64; Ω_2=288; Ω_3=392; Pr=1,294* 10^{-5}.

There are 1419 six-storey phase trees. 95 of them are phase trees of range four. Only four of them are elementary EP(4,6).

ELEMENTARY PARTICLES EP(5,7) OF RANGE FIVE

{...} —cipher code. Ω_0, Ω_1, Ω_2, Ω_3, Ω_4 —parts of the inflation potential Ω.
Pr —probability of direct emergence.

{19,7,47,55,59,19,125,21,57,17,126,46,22,58,18,124,20,56,16}
Ω = 41953; Ω_0=92; Ω_1=115; Ω_2=1657; Ω_3=30323; Ω_4=9766; Pr = $4,475*10^{-8}$

{19,7,47,23,123,19,125,21,57,17,126,46,22,58,18,124,20,56,16}
Ω = 34165, Ω_0=92; Ω_1=117; Ω_2=1411; Ω_3=24448; Ω_4=8097; Pr = $6.821*10^{-8}$

{18,7,47,55,59,19,125,21,57,17,126,54,58,18,124,20,56,16}
Ω = 37387, Ω_0=92; Ω_1=116; Ω_2=1627; Ω_3=27207; Ω_4= 83345;
Pr = $4.475*10^{-8}$

{17,7,55,123,19,125,21,57,17,126,54,58,18,124,20,56,16}
Ω = 26878, Ω_0=92; Ω_1=119; Ω_2=1134; Ω_3=19065; Ω_4=6468;
Pr = $1.220*10^{-7}$

{19, 7,47,55,59,19,125,45,57,17,126,46,22,58,18,124,44,56,16}
Ω = 30948, Ω_0=92; Ω_1=115; Ω_2=1649; Ω_3= 23150; Ω_4=5942;
Pr = $5.214*10^{-8}$

{19,7,47,23,123,19,125,45,57,17,126,46,22,58,18,124,44,56,16}
$\Omega = 26646$, $\Omega_0=92$; $\Omega_1=117$; $\Omega_2=1403$; $\Omega_3=19644$; $\Omega_4=5390$;
$Pr = 7.634*10^{-8}$

{21,7,47,119,123,43,125,45,21,121,41,126,46,54,122,42,124,44,20,120,40}
$\Omega > 4*10^5$, $Pr = 5.127*10^{-9}$,

{22,7,47,55,123,43,19,125,45,21,121,41,126,46,54,122,42,124,44,20,120,40}
$\Omega > 4*10^5$, $Pr = 1.084*10^{-8}$.

There are nearly a half million seven-storey phase trees. Only eight of them are elementary EP(5,7).

There are millions of eight-storey phase trees and only 21 of them are elementary EP(6,8) of range six.. There are millions of nine-storey phase trees and 86 of them are elementary EP(7,9) of range 7.

In Figure 1 we present a simple method to construct EP(R,R+2) phase trees. We can utilize phase tree EP(3,5) ={5,5,11,14,4} as right part of the phase tree EP(4,6). Phase tree EP(3,5) has only one acceptable free dot to hang up element /\ and to provide left part of the EP(4,6). Phase tree EP(4,6) has two acceptable free dots to hang up elements /\ and to construct two EP(5,7) et cetera. We have only three EP(3,5) to construct all EP(R,R+2) by this method. The clue is to find acceptable free dots. First Results are on Figures 2–5. One can prolong construction of EP(R,R+2) to extend classification. For EP(R,R+2)

$$\Omega_0(R) = 2\,\Omega_0(R-1) + 4. \quad (2)$$

Each phase tree has special cipher code. First number of code is number of chains plus 1. It is length of the code to use when we create programs for computer. Second number is height of the tree –number of edges in the first chain. After that follow codes of chains. Code of chain is integer. When written in binary system it tells that chain must turn to left (cipher 1) or to right (cipher 0).

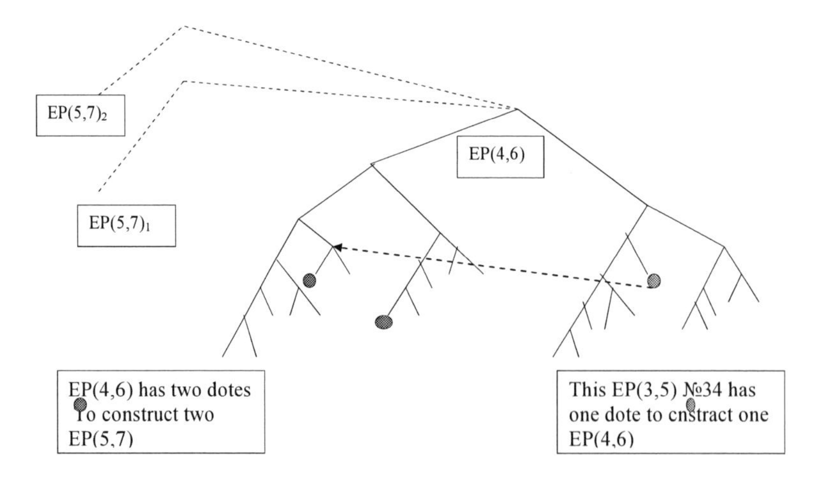

Figure 1. Method to construct elementary phase trees EP(R,R+2) (example)

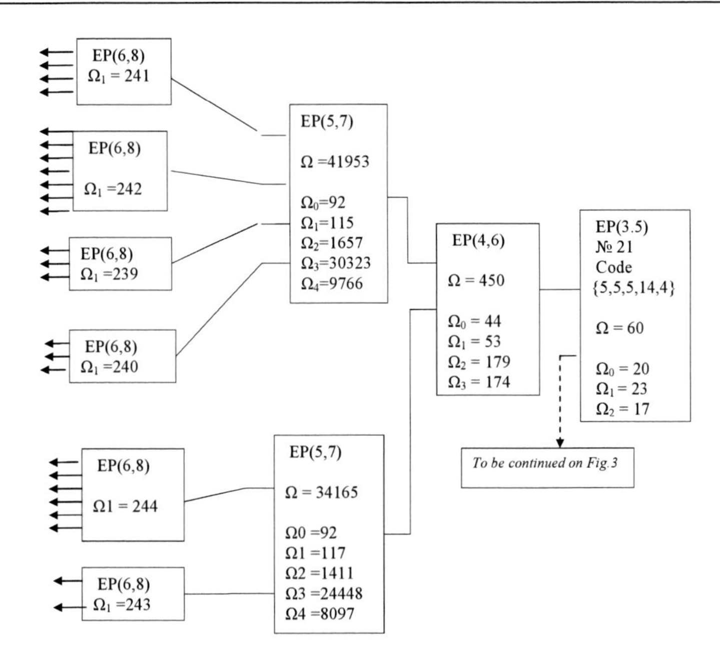

Figure 2. Construction of the phase trees EP(R,R+2) from EP(3,5) (syntax No 21).

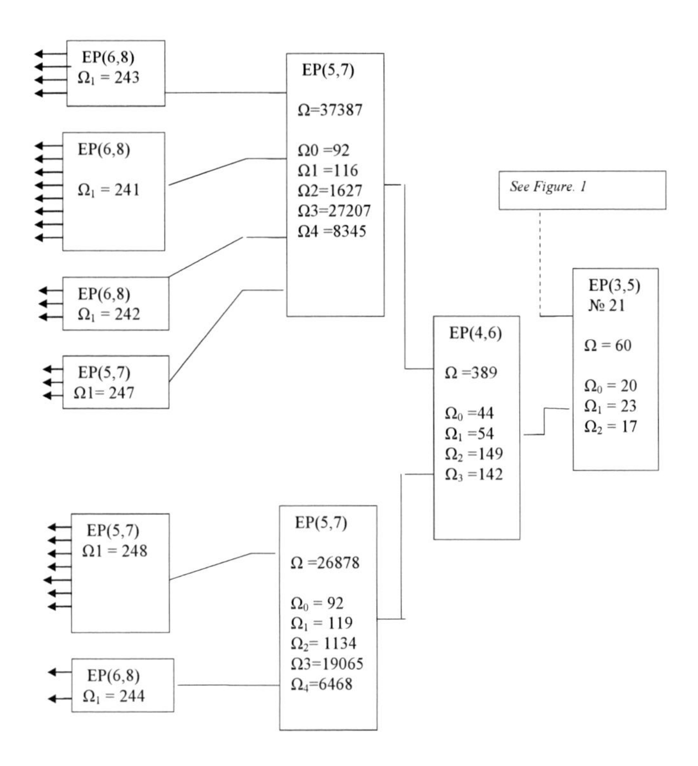

Figure 3. Construction of the phase trees EP(R,R+2) from EP(3,5) (syntax No 21).

Additionally to EP(R,R+2) there are many elementary phase trees EP(R, R+n) (n>2) with small probability to arise. So phase tree {5,5,5,14,4} belongs to the phases with codes ($H \geq 5$):

$$\{5, H, (3*2^{H-4}-1), (2^{H-1}-2), (3*2^{H-4}-2)\}.$$

Potentials are: $\Omega = (3H^2+13H)/2 - 10$; $\Omega_0 = 4H$; $\Omega_1=H^2-H+3$; $\Omega_2=(H^2+7H)/2 - 13$.
Phase tree {5,5,11,14,4} belongs to the phases EP(3,H) with codes

$$\{5, H, (3*2^{H-3}-1\}, (2^{H-1}-2), (3*2^{H-4}-2)\}.$$

Potentials are: $\Omega = (3H^2+9H)/2 - 4$; $\Omega_0=4H$; $\Omega_1=H^2-H+3$; $\Omega_2=(H^2+3H)/2-7$.
Phase tree {6,5, 11,5,30,10} belongs to the phase trees EP(3,H) with codes

$$\{6, H, (3*2^{H-3}-1), (3*2^{H-4}-1), (2^{H}-2), (3*2^{H-3}-2)\}.$$

Potentials are: $\Omega = (3H^2+23H)/2 - 11$; $\Omega_0=4(H+1)$; $\Omega_1=H^2+3$; $\Omega_2=(H^2+15H)/2-18$.
Probabilities to arise fell nearly by factor 10 at growing from H to H+2.

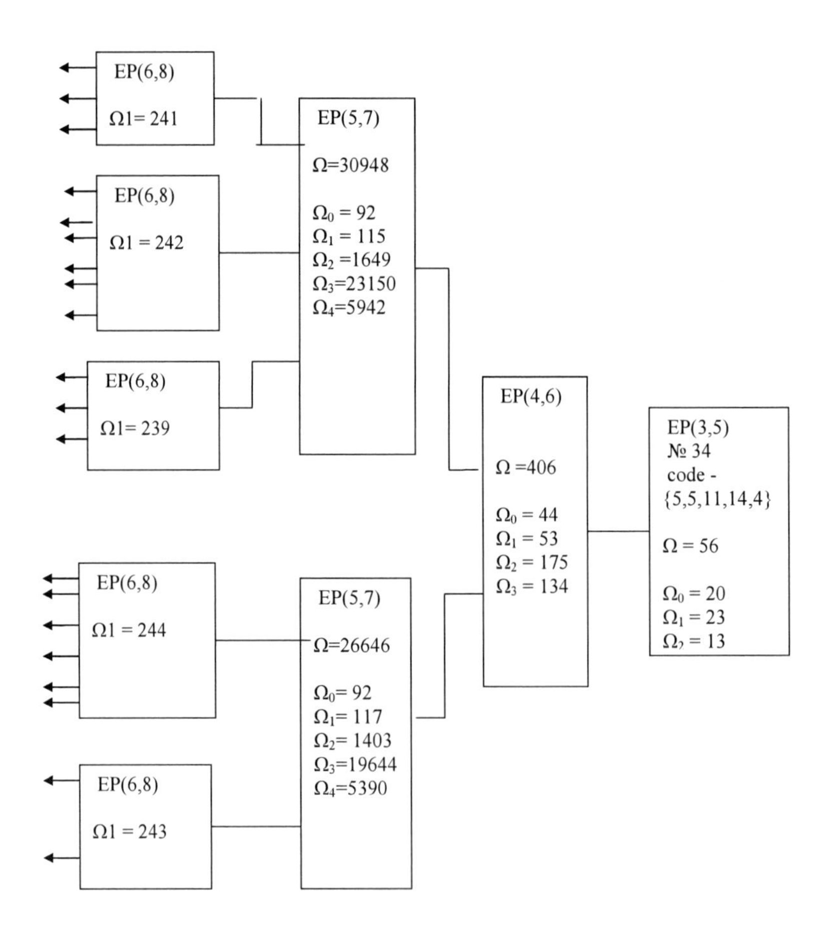

Figure 4. Construction of EP(R,R+2) from EP(3,5) syntax № 34.

Reciprocally we can classify phase trees EP(4,H) ($H \geq 6$):

$\{10,H,(3*2^{H-3}-1),(3*2^{H-4}-1),(15*2^{H-5}-1),(5*2^{H-5}-1),(2^{H}-2),(3*2^{H-4}-),$
$(15*2^{H-5}-2),(5*2^{H-5}-2)\}$
$\Omega(H) = (3H^2+135H)/2 - 9$
$\Omega_0=4H+20$; $\Omega_1=H^2+17$; $\Omega_2=(H^2+113H)/2-178$; $\Omega_3=7H+132$;

$\{9,H,(7*2^{H-4}-1),(15*2^{H-5}-1),(5*2^{H-5}-1),(2^{H}-2),(3*2^{H-4}-2),(15*2^{H-5}-2),$
$(5*2^{H-5}-2)\}$
$\Omega(H) = (3H^2+115H)/2 - 10$
$\Omega_0=4H+20$; $\Omega_1=H^2+18$; $\Omega_2=(H^2+97H)/2-160$; $\Omega_3=5H+112$;

$\{10,H,(3*2^{H-3}-1),(3*2^{H-4}-1),(15*2^{H-5}-1),(5*2^{H-5}-1),(2^{H}-2),(3*2^{H-3}-),$
$(15*2^{H-5}-2),(5*2^{H-5}-2)\}$
$(H) = (3H^2+127H)/2 - 29$
$\Omega_0=4H+20$; $\Omega_1=H^2+17$; $\Omega_2=(H^2+105H)/2-158$; $\Omega_3=7H+92$;

$\{11,H,(3*2^{H-3}-1),(7*2^{H-4}-1),(31*2^{H-5}-1),(11*2^{H-5}-1),(2^{H}-2),(3*2^{H-3}-2),$
$(3*2^{H-4}-2),(31*2^{H-5}-2),(11*2^{H-5}-2)\}$
$\Omega(H) = (3H^2+205H)/2 +127$
$\Omega_0=4H+28$; $\Omega_1=H^2+28$; $\Omega_2=(H^2+187H)/2-291$; $\Omega_3=5H+362$;

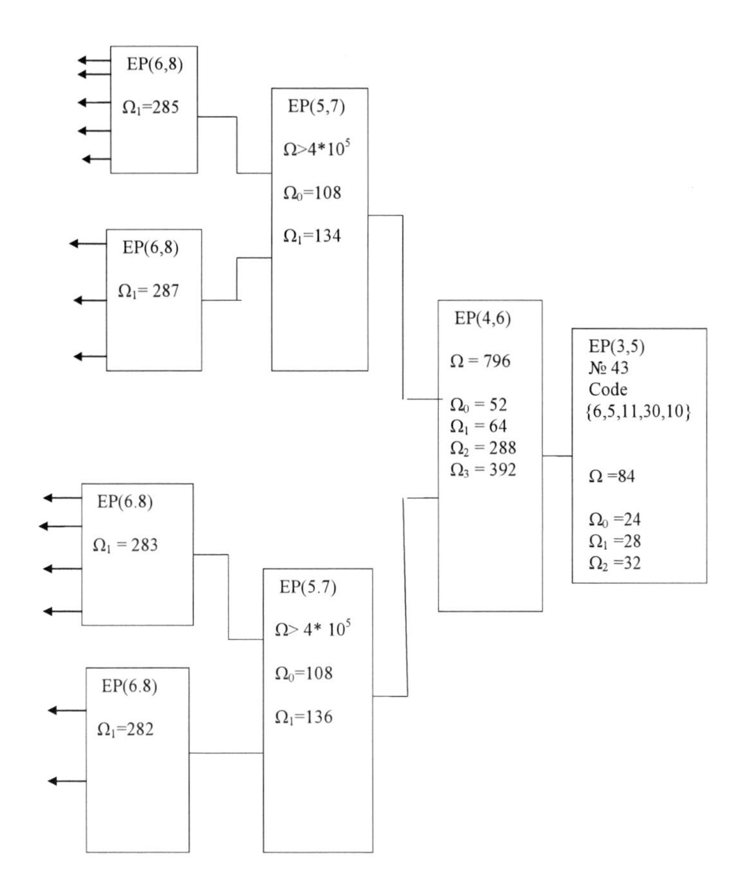

Figure 5. Construction of EP(R,R+2) from EP(3,5) (syntax № 43).

Elementary Eight-Storey Phase Trees of Range Six

From EP(5,7)-EP(4,6) – EP(3,5) 1-1-21

{ 37, 8, 95, 239, 119, 39, 251, 43, 115, 35, 253, 93, 45, 117, 37, 249, 41, 113, 33, 254, 94, 110, 118, 38, 250, 42, 114, 34, 252, 92, 44, 116, 36, 248, 40, 112, 32 }
Ω0 = 188; Ω1=241; 1

{ 37, 8, 95, 111, 247, 39, 251, 43, 115, 35, 253, 93, 45, 117, 37, 249, 41, 113, 33, 254, 94, 110, 118, 38, 250, 42, 114, 34, 252, 92, 44, 116, 36, 248, 40, 112, 32 }
Ω0 = 188; Ω1=242; 2

{ 38, 8, 95, 111, 119, 39, 251, 91, 43, 115, 35, 253, 93, 45, 117, 37, 249, 41, 113, 33, 254, 94, 110, 118, 38, 250, 42, 114, 34, 252, 92, 44, 116, 36, 248, 40, 112, 32 }
Ω0 = 188; Ω1=239; 3

{ 37, 8, 95, 111, 119, 39, 251, 107, 115, 35, 253, 93, 45, 117, 37, 249, 41, 113, 33, 254, 94, 110, 118, 38, 250, 42, 114, 34, 252, 92, 44, 116, 36, 248, 40, 112, 32 }
Ω0 = 188; Ω1=240; 4

2 – 1 - 21

{ 37, 8, 95, 111, 247, 39, 251, 43, 115, 35 253, 93, 45, 117, 37, 249, 41, 113, 33, 254, 94, 46, 246, 38, 250, 42, 114, 34, 252, 92, 44, 116, 36, 248, 40, 112, 32 }
Ω0 = 188; Ω1=244; 1

{ 38, 8, 95, 47, 247, 39, 251, 91, 43, 115, 35 253, 93, 45, 117, 37, 249, 41, 113, 33, 254, 94, 46, 246, 38, 250, 42, 114, 34, 252, 92, 44, 116, 36, 248, 40, 112, 32 }
Ω0 = 188; Ω1=243; 2

1 – 2 - 21

{ 35, 8, 95, 239, 119, 39, 251, 43, 115, 35, 253, 109, 117, 37, 249, 41, 113, 33, 254, 94, 110, 118, 38, 250, 42, 114, 34, 252, 108, 116, 36, 248, 40, 112, 32 }
Ω0 = 188; Ω1 = 243; 1

{ 35, 8, 95, 111, 247, 39, 251, 43, 115, 35, 253, 109, 117, 37, 249, 41, 113, 33, 254, 94, 110, 118, 38, 250, 42, 114, 34, 252, 108, 116, 36, 248, 40, 112, 32 }
Ω0 = 188; Ω1 = 244 2

{ 36, 8, 95, 111, 119, 39, 251, 91, 43, 115, 35, 253, 109, 117, 37, 249, 41, 113, 33, 254, 94, 110, 118, 38, 250, 42, 114, 34, 252, 108, 116, 36, 248, 40, 112, 32 }
Ω0 = 188; Ω1 = 241 3

{ 35, 8, 95, 111, 119, 39, 251, 107, 115, 35, 253, 109, 117, 37, 249, 41, 113, 33, 254, 94, 110, 118, 38, 250, 42, 114, 34, 252, 108, 116, 36, 248, 40, 112, 32 }
Ω0 = 188; Ω1 = 242 4

2 – 2 - 21

{ 34, 8, 95, 111, 247, 39, 251, 43, 115, 35, 253, 109, 117, 37, 249, 41, 113, 33, 254, 110, 246, 38, 250, 42, 114, 34, 252, 108, 116, 36, 248, 40, 112, 32 }
Ω0 = 188; Ω1 = 247 1

{ 33, 8, 111, 247, 39, 251, 107, 115, 35, 253, 109, 117, 37, 249, 41, 113, 33, 254, 110, 246, 38, 250, 42, 114, 34, 252, 108, 116, 36, 248, 40, 112, 32 } 2

1 – 1 – 34

{ 37, 8, 95, 239, 119, 39, 251, 91, 115, 35, 253, 93, 45, 117, 37, 249, 89. 113, 33, 254, 94, 110, 118, 38, 250, 90, 114, 34, 252, 92, 44, 116, 36, 248, 88, 112, 32 }
Ω0 = 188; Ω1 = 241 1

{ 37, 8, 95, 111, 247, 39, 251, 91, 115, 35, 253, 93, 45, 117, 37, 249, 89. 113, 33, 254, 94, 110, 118, 38, 250, 90, 114, 34, 252, 92, 44, 116, 36, 248, 88, 112, 32 }
Ω0 = 188; Ω1 = 242 2

{ 38, 8, 95, 111, 119, 39, 251, 91, 43, 115, 35, 253, 93, 45, 117, 37, 249, 89. 113, 33, 254, 94, 110, 118, 38, 250, 90, 114, 34, 252, 92, 44, 116, 36, 248, 88, 112, 32 }
Ω0 = 188; Ω1 = 239 3

2 – 1 - 34

{ 37, 8, 95, 111, 247, 39, 251, 91, 115, 35, 253, 93, 45, 117, 37, 249, 89, 113, 33, 254, 94, 46, 246, 38, 250, 90, 114, 34, 252, 92, 44, 116, 36, 248, 88, 112, 32 }
Ω0 = 188; Ω1 = 244 1

{ 38, 8, 95, 47, 247, 39, 251, 91, 43, 115, 35, 253, 93, 45, 117, 37, 249, 89, 113, 33, 254, 94, 46, 246, 38, 250, 90, 114, 34, 252, 92, 44, 116, 36, 248, 88, 112, 32 }
Ω0 = 188; Ω1 = 243 2

1 – 1 - 43

{ 42, 8, 95, 239, 247, 87, 39, 251, 91, 43, 243, 83, 253, 93, 109, 245,
85, 249, 89, 41, 241, 81, 254, 94, 238, 246, 86, 250, 90, 42, 242, 82,
252, 92, 108, 244, 84, 248, 88, 40, 240, 80]
Ω0 = 220 Ω1 = 285 1

{ 41, 8, 95, 239, 247, 87, 251, 91, 107, 243, 83, 253, 93, 109, 245, 85, 249, 89, 41, 241,
81, 254, 94, 238, 246, 86, 250, 90, 42, 242, 82, 252,
92, 108, 244, 84, 248, 88, 40, 240, 80]
Ω0 = 220 Ω1 = 2

2 – 1 - 43

{ 43, 8, 95, 239, 247, 87, 39, 251, 91, 43, 243, 83, 253, 93, 109, 245, 85, 249, 89, 41, 241,
81, 241, 94, 110, 246, 86, 38, 250, 90, 42, 242, 82, 252, 92,108, 244, 84, 248, 88, 40 , 88,
40, 240, 80 }
Ω0 = 220 Ω1 = 286 1

{ 43, 8, 95, 111, 247, 87, 39, 251, 91, 107, 243, 83, 253, 93, 109, 245, 85, 249, 89, 41,
241, 81, 2 1, 241, 94, 110, 246, 86, 38, 250, 90, 42, 242, 82, 252, 92,108, 244, 84, 248,
88, 40 , 88, 40, 240, 80 }
Ω0 = 220 Ω1 = 282 2

In the process of imitation by computer, there commonly arise phases of inflation which are not elementary P(R,H). They are excited or compound particles. Excited elementary particles EP(R,H) interact if they have common subtrees. Subtrees of range r < R play part of the force carriers. They are derivatives of the inflation potential Ω. Each elementary particle can be carrier of the force (supersymmetry). At each scattering there arise many new subtrees. Frequency of them is one, One can say, that they are fermions. At following steps of inflation, some of them multiplicate end they transmute into bosons. Supersymmetry breaks.

Frequencies of the phase trees of range 0–3 soon become very high. One can say that in this theory three-dimensions space is bosonic condensate and superfluid. It freely penetrates all objects. After more scattering new long living fermions of high range emerge. They would be particles to collate with standard model. To attain this high range epoch of long living fermions one needs a big computer.

Phase trees have different probabilities to come about. Among 4, 5 and 6-storey phase trees, high probabilities exceed low probabilities by factor 8, 305 and 1782 respectively. In Figure 6, we present probabilities of direct emergence for the first 300 from 1419 six-storey phase trees (in order of syntaxis numbers). Therefore, commonly the process of inflation will create most probable particles. Each phase tree is a unique scenario of the process. One can say that each phase tree is a unique universe and that physical reality is a collection of scenarios in our mind. But it is not multiverse of H. Everett or of A. Linde. All phase trees have many common low-range eigensubtrees — physical reality. They all have the same elementary particles and nearly the same three-dimensional space. After following high-range evolution, phase trees can become very unique and isolated objects with high organization.

References

[1] AB. Khromov. Measurements in a special chaotic system. Moscow. Measuring,48, 3, 2004.

[2] AW. Khromov Modelling inflation. Moscow, nonlinear world, № 3, Volume 4, 2006. p.104, ed. Radioelectronics.

[3] AB. Khromov, AA. Kovalev Quantum system observables. Moscow, Metrology, 2006, No. 6, 3-12.

[4] AF; Kotyukov, AB. Khromov Prospects of quantum metrology. On the basis of physical "theory of everything" (Theory of everything). Moscow, Metrology No. 12, 2006, 3-12 .

[5] AF Kotyukov, AV Khromov On the possible origin and computability of physical constants. Moscow train. Metrology № 4 2008, 3-12.

[6] AF Kotyukov, AB. Khromov Theoretical Foundations inflationary quantum metrology. Moscow. Metrology, 2008, No. 8, 3-30. http:// www.springlink.comopenurl. asp?

In: Superstring Theory in the 21st Century...
Editor: G. B. Charney, pp.13-86
ISBN 978-1-61668-385-6

Chapter 2

MODULAR FORMS AND SUPERSTRINGS AMPLITUDES

***L. Sergio Cacciatori*[1]* *and Dalla Francesco Piazza*[2] †**
[1]Dipartimento di Fisica e Matematica, Universitȧ dell'Insubria Milano, 22100 Como, Italy, and I.N.F.N., sezione di Milano, Italy.
[2] Dipartimento di Fisica e Matematica, Universitȧ dell'Insubria Milano, 22100 Como, Italy, and I.N.F.N., sezione di Milano, Italy

Abstract

We will present the recent developments in constructing superstring amplitudes for genus g higher than 2.

The main problem in defining superstring amplitudes is due to the ambiguity of the super covariant path integral formulation. Super Weyl and super diffeomorphism invariance of the theory give rise to certain constraints on the vacuum-to-vacuum amplitudes, which must then behave as suitable modular forms. A strategy for determining such amplitudes is then to impose physical constraints to opportune subspaces of modular forms selected by the above symmetries. Recently, this method has been applied to determine the vacuum-to-vacuum amplitudes for superstrings at genus three and four.

Here we will explain the mathematical and physical details underlying such constructions, with particular attention to the $g = 3$ case. Differently from the bosonic case, it will be shown that the main ingredients are certain equivariant modular forms supporting finite groups having a physical meaning. The physical constraints can then be recasted in terms of representation theory and the problem reduces itself in selecting the invariants.

We will show that the constraints together with some simple assumptions give rise to a unique solution. However, we also provide some criticism, evidencing some related open problems.

*E-mail address: sergio.cacciatori@uninsubria.it
†E-mail address: f.dallapiazza@uninsubria.it

1. Introduction

The aim of these notes is to provide an exposition as of the very recent developments in solving the problem of find a consistent candidate for the superstring measure. In particular we will concentrate on our papers, written in collaboration with Bert Van Geemen, [CD, CDG1, CDG2], and will try to be as explicit as possible, presenting many inedit details, in order to make all results easily reproducible and to clarify many subtle points. Let us begin with an adequate presentation of the problem.

1.1. The Bosonic History

To understand the strategy underling the construction of vacuum-to-vacuum superstring amplitudes in the NRS formalism, we think it is a good choice to start with the main rigorous results for the bosonic case. For simplicity we will work only with closed strings.

In flat Minkowski space the Polyakov action on a Riemann surface Σ_g of genus g and with metric h is

$$\begin{aligned} I_g(X,h) &= \frac{1}{4\pi\alpha'}\int_{\Sigma_g} d^2z\sqrt{h}\, h^{ab}\partial_a X\cdot\partial_b X, \\ X &: \Sigma_g \hookrightarrow \mathbb{R}^D. \end{aligned}$$

It formally selects the weight measure for the path integral formulation of the bosonic partition function

$$Z^g_{bos} = \int [Dh_{ab}][DX]\exp(-I(X,h)), \tag{1}$$

where the functional sums are performed over all possible metrics over Σ_g and over all maps $X : \Sigma_g \hookrightarrow \mathbb{R}^D$. The whole partition function obviously involves a sum over all genera. When the path integral is well defined, one can extend it to compute amplitudes involving vertex operators for example momenta K_i^μ of mass $-m_i^2 = K_i\cdot K_i$

$$\left\langle \prod_{i=1}^N V_i(K_i)\right\rangle = \int [Dh_{mn}][DX]\exp(-I(X,h))\prod_{i=1}^N V_i(K_i),$$

where, for example,

$$\begin{aligned} V_{-1}(K) &= \int d^2z\sqrt{h}\, e^{iK\cdot X}, && \text{for the Tachyon,} \\ V_0(K) &= \int d^2z\sqrt{h}\,\partial_a X\cdot\partial^a X e^{iK\cdot X}, && \text{for the Graviton.} \end{aligned}$$

To define the path integral the main idea is to make use of the very large symmetry group of the classical theory that is the semidirect product $G = \mathrm{Weyl}(\Sigma_g)\ltimes \mathrm{Diff}(\Sigma_g)$ of the group of Weyl transformations times the group of diffeomorphisms of the Riemann surface Σ_g. If M is the set of all possible Riemannian metrics over Σ_g, then the moduli space for conformal classes of Riemann surfaces is $\mathcal{M}_g = M/G$. It is a finite dimensional complex manifold with holomorphic dimension

$$\dim_{\mathbb{C}}\mathcal{M}_g = \begin{cases} 0 & \text{if} \quad g = 0 \\ 1 & \text{if} \quad g = 1 \\ 3g-3 & \text{if} \quad g \geq 2. \end{cases}$$

At infinitesimal level, diffeomorphisms can be thought as locally generated by vector fields v and scalar fields ω:

$$\begin{array}{lcr} \text{Diff}(\Sigma_g) & : & \delta h_{ab} = \nabla_a v_b + \nabla_b v_a; \\ \text{Weyl}(\Sigma_g) & : & \delta h_{ab} = 2\omega h_{ab}. \end{array}$$

One expects to be able to reduce the path integral to a finite dimensional integral over $\mathcal{M}_g$. Indeed, the functional summation over the X fields can be easily performed being the action a quadratic functional of X. The Gaussian integration is then expressed in terms of the determinant of the Laplacian associated to the given metric h_{ab} over Σ_g, and which can be computed, for example, by means of a zeta function regularization. However, in computing the determinant one have to be careful about the existence of zero modes which must dropped out. Such regularization breaks the conformal invariance making the procedure anomalous so that the machinery fails to work. However, the anomaly disappears when the target space dimension is $D = 26$ and the path integral is hopefully well defined. This is a well known result which can be obtained in many ways and we will not review it here. We will rather look at the approach evidencing the complex geometry underling the Riemann surfaces in place of the spectral properties of the related Laplacian. This is indeed the approach introduced by Belavin and Knizhnik [BK], Beilinson and Manin [BM] and adopted by d'Hoker and Phong [DP7] in order to finally solve the problem of computing two loop superstring amplitudes.

1.1.1. Path Integral and Complex Geometry

Let us now collect here some technical points which permits to reexpress the path integral formula in terms of some geometric data, suitable for obtaining the subsequent main theorems, see [DP7].

We have to evaluate the path integral where the measure is defined by the metrics

$$\begin{aligned} \|\delta h\|^2 &= \int_{\Sigma_g} \sqrt{\det h} h^{ab} h^{cd} \delta h_{ac} \delta h_{bd} d^2\zeta, \\ \|\delta X\|^2 &= \int_{\Sigma_g} \sqrt{\det h} \delta X \cdot \delta X d^2\zeta. \end{aligned}$$

To sum over inequivalent metric configurations we need to specify some coordinates over the moduli space. This means that we need to distinguish between symmetry transformations (diffeomorphisms and conformal transformations) and genuine transformations. A particular interesting choice is given by introducing isothermal coordinates, which are such that $ds^2 = 2h_{z\bar{z}}dzd\bar{z}$. Recall that to a given metric it is associated a complex structure

$J_a^{\ b} = h^{\frac{1}{2}}\varepsilon_{ac}h^{ac}$, so that isothermal coordinates are determined by solving the Beltrami's equations

$$J_a^{\ b}\frac{\partial z}{\partial \zeta^b} = i\frac{\partial z}{\partial \zeta^a}.$$

Deformations of the complex structure are parameterized with Beltrami differentials $\mu_{i\bar{z}}^{\ z}$, $i = 1, 2, \ldots, 3g-3$ so that

$$\delta h_{\bar{z}\bar{z}} = \sum_{i=1}^{3g-3} t^i h_{z\bar{z}}\mu_{i\bar{z}}^{\ z}, \tag{2}$$

for certain complex parameters t^i. It follows that two deformations μ, $\tilde{\mu}$ are equivalent in $\mathcal{M}_g$ if $\tilde{\mu}_{\bar{z}}^{\ z} - \mu_{\bar{z}}^{\ z} = \partial_{\bar{z}}v^z$ for some vector field v^z. It then results that the tangent moduli space is

$$T\mathcal{M}_g = \{\text{Beltrami differentials } \mu_{i\bar{z}}^{\ z}\}/\{\text{Range of } \partial_{\bar{z}} \text{ over vector fields }\}$$

The cotangent bundle is parameterized by quadratic differentials Φ_{zz} which are non singularly paired to Beltrami differentials by

$$\langle\mu|\Phi\rangle = \int d^2\mu_{\bar{z}}^{\ z}\Phi_{zz}.$$

It is not difficult to see from this that the quadratic differentials are orthogonal to the G-transformations in the following sense. Let be (ω, v) a generator of an infinitesimal symmetry transformation: if the metric is $h = 2e^{2\phi}dzd\bar{z}$, then

$$\delta_{(\omega,v)}h = 2e^{2\phi}\bar{\partial}v(d\bar{z})^2 + 2e^{2\phi}\partial\bar{v}(dz)^2 + 2[2\omega e^{2\phi} + \partial(e^{2\phi}v) + \bar{\partial}(e^{2\phi}\bar{v})]dzd\bar{z}.$$

Note that the squared norm of this variation is

$$\begin{aligned}\|\delta_{(\omega,v)}h\|^2 &= \int_{\Sigma_g} 2e^{2\phi}(2\omega + e^{-2\phi}\partial(e^{2\phi}v) + e^{-2\phi}\bar{\partial}(e^{2\phi}\bar{v}))^2 d^2\zeta \\ &+ \int_{\Sigma_g} 2e^{2\phi}(\partial\bar{v})(\bar{\partial}v)d^2\zeta,\end{aligned}$$

from which one easily computes the Jacobian of the transformation. It then follows that the metric deformations orthogonal to $\delta_{(\omega,v)}h$ are the ones of the form

$$\delta_\perp h = \Phi_{zz} + \bar{\Phi}_{\bar{z}\bar{z}}$$

where Φ are the holomorphic quadratic differentials ($\bar{\partial}\Phi = 0$).
The space of holomorphic quadratic differentials is a particular case of the space V^n of holomorphic n-differentials, holomorphic covariant tensors of rank n, for n positive, end contravariant of rank $-n$ for negative n. V^n is naturally provided with the hermitian product

$$\langle\tau,\tau'\rangle_n = \int_{\Sigma_g}(e^{2\phi})^{1-n}\bar{\tau}\tau' d^2\zeta.$$

Moreover, on V^n it acts the n-Laplacian

$$\Delta_n \tau = -\frac{1}{2}e^{2(n-1)\phi}\partial(e^{-2n\phi}\partial\tau).$$

A choice of a basis μ^i, $i = 1, \ldots, 3g-3$, of Beltrami differentials, so that the metric, according to Beltrami, takes the form $h = 2e^{2\phi}|dz + \sum_i t^i \mu_i d\bar{z}|^2$, determine a dual basis Φ^j of V^2 such that $\langle \Phi^j, \mu_i \rangle = \delta^i_j$. Then, the matrix $G^{ij} = \langle \Phi^i, \Phi^j \rangle_2$ is invertible with inverse G_{ij} and

$$\mu_i = \frac{1}{2}e^{-2\phi}G_{ij}\bar{\Phi}^j.$$

Then we finally get

$$||\delta_\perp h||^2 = G_{ij}\delta t^i \delta\bar{t}^j,$$

and we are able to compute the Jacobian transformation

$$\begin{aligned}[Dh_{ab}] &= \det\Delta_{-1}[D\omega][Dv]|\det\langle\mu_i|\Phi^j\rangle|^2 \det \\ &\cdot G_{ij}(1/2)^{3g-3}dt^1 \wedge \ldots \wedge dt^{3g-3} \wedge d\bar{t}^1 \wedge \ldots \wedge d\bar{t}^{3g-3}.\end{aligned}$$

Dropping the gauge volume given by integration over ω and v, and performing the Gaussian integral[1] in X, we obtain the desired expression for the bosonic partition function

$$Z^g_{BOS} = \int d^{3g-3}t\, d^{3g-3}\bar{t}\, \frac{|\det\langle\mu_i|\Phi^j\rangle|^2}{\det\langle\Phi^i|\Phi^j\rangle} \det\Delta_{-1}\left(\frac{\det\Delta_0}{\int h_{z\bar{z}}d^2\zeta}\right)^{-\frac{D}{2}}. \tag{3}$$

It is also possible to give a gauge invariant formulation, adding ghosts $b = b_{zz}dz^2$, $c = c^z dz^{-1}$, $I_{gh} = \frac{1}{2\pi}\int d^2z(b_{zz}\partial_{\bar{z}}c^z + \bar{b}_{\bar{z}\bar{z}}\partial_z\bar{c}^{\bar{z}})$ which satisfy

$$\int [D(b\bar{b}c\bar{c})]e^{-I_{gh}}\left|\prod_{i=1}^{3g-3}\langle\mu_i|b\rangle\right|^2 = \det(\bar{\partial}_2^\dagger\bar{\partial}_2)\frac{|\det\langle\mu_i|\Phi_j\rangle|^2}{\det\langle\Phi_i|\Phi_j\rangle}$$

and then

$$Z^g_{BOS} = \int [D(b\bar{b}c\bar{c}X)]\prod_{i=1}^{3g-3}|\langle\mu_i|b\rangle|^2 e^{-(I_g+I_{gh})}. \tag{4}$$

The expression (**??**) is generically anomalous, presenting two kinds of anomalies. The first one is crucial for defining the theory and is the conformal anomaly, which breaks invariance of Z^g_{BOS} under Weyl transformations. By means of Heat-Kernel methods one can show that[2], if $\Phi^{(n)} \in V^n$,

$$\delta_\omega \log \frac{\det\Delta_n}{\det\langle\Phi_i^{(n)}|\Phi_j^{(n)}\rangle\det\langle\Phi_i^{(1-n)}|\Phi_j^{(1-n)}\rangle} = -\frac{6n^2-6n+1}{6\pi}\int_{\Sigma_g} d^2z\sqrt{h}R\omega,$$

[1]Which gives $(\det'\Delta_0/\int\sqrt{h}d^2\zeta)^{-\frac{D}{2}}$, where the prime means that zero modes do not contribute to the determinant, but contribute to the volume factor.

[2]To be more precise, a diverging additive term depending only on the worldsheet volume appears, but it is innocuous and can be adsorbed in a constant counterterm in the bosonic string action.

where R is the worldsheet scalar curvature. Using this in $\delta_\omega Z_{BOS}$ we see that the conformal anomaly vanish if $D = 26$. Thus, conformal invariance constraints the theory to work with a 26 dimensional space (at least in the case when the target space is the flat Minkowski space).
The second anomaly is the so called Holomorphic anomaly. The Beltrami equation says that the operator $\partial_{\bar{z}}$ depends holomorphically on the moduli: $\partial_{\bar{z}} \to \partial_{\bar{z}} - \mu_{\bar{z}}^{\ z}\partial_z$. Now, expression (??) makes evident that Z^g_{BOS} should be expected to behave as the square modulus of an holomorphic function. An obstacle for this to happen comes from the Belavin-Knizhkin theorem:

$$\delta_{\bar{\mu}}\delta_\mu \log \frac{\det(\bar{\partial}_n^\dagger\bar{\partial}_n)}{\det\langle\Phi_i^{(n)}|\Phi_j^{(n)}\rangle \det\langle\Phi_i^{(1-n)}|\Phi_j^{(1-n)}\rangle}$$
$$= -\frac{6n^2-6n+1}{6\pi}\int_\Sigma d^2z\nabla_{\bar{z}}\mu\nabla_z\bar{\mu}\omega.$$

Note however that, exactly as for the conformal anomaly, the holomorphic anomaly vanishes if $D = 26$. This fortunate coincidence (if this is the case) permits to eliminate both anomalies simultaneously. The importance to cancel the holomorphic anomaly for bosonic strings is not an evident fact as, however it is an advantageous fact because as it permits to reexpress the partition function in terms of global objects, as it has been shown in the remarkable paper of Belavin and Knizhkin [BK], and in a work of Beilinson and Manin [BM] and Manin [MYu1, MYu2].
A starting point is a theorem of Mumford [Mu2] which was able to proof that the linear bundle $U = K \otimes \lambda^{-13}$ is a holomorphically trivial bundle over $\mathcal{M}_g$. Here K is the canonical bundle over $\mathcal{M}_g$, that is the highest wedge power of the cotangent bundle. A local basis is $\Phi_1 \wedge \ldots \wedge \Phi_{3g-3}$. Similarly, λ is the Hodge bundle over $\mathcal{M}_g$, the highest wedge power of the holomorphic cotangent bundle, which is generated by the Abelian differentials $\omega_1 \wedge \ldots \wedge \omega_g$. As a consequence, U admits an essentially unique global holomorphic section ψ_g, the Mumford section. Expressed in terms of a local basis the Mumford theorem says that it exists a unique holomorphic function over $\mathcal{M}_g$ such that

$$\psi_g = F\frac{\Phi^1 \wedge \ldots \wedge \Phi^{3g-3}}{(\omega_1 \wedge \ldots \wedge \omega_g)^{13}}$$

is a global section of U. Moreover, ψ_g is nonvanishing everywhere, and meromorphic at infinity with an order two pole.
The Belavin-Knizhkin theorem implies that the bosonic partition function, apart from a constant, is given by the square modulus of the Mumford section:

$$Z_{BOS} = c_g \int |F|^2(-i)^g\Phi_1 \wedge \ldots \wedge \Phi_{3g-3} \wedge \bar{\Phi}_1 \wedge \ldots \wedge \bar{\Phi}_{3g-3}| \det \int_{\Sigma_g} \bar{\omega}_I \wedge \omega_J|^{-13}. \quad (5)$$

The Mumford theorem has been proved by Beilinson and Manin [BM] in a strongest form which permits to be more explicit. Moreover, Manin [MYu1, MYu2] has been able to provide explicit expressions in terms of theta functions. Recall that at a canonical basis A_I,

B_I, $I = 1, 2, \ldots, g$ for the homology of Σ_g, $A_I \cap A_J = B_I \cap B_J = 0$, $A_I \cap B_J = \delta_{IJ}$, one can associate a canonical basis ω_I for the Abelian differentials by

$$\int_{A_I} \omega_J = \delta_{IJ}.$$

Then

$$\Omega_{IJ} := \int_{B_I} \omega_J$$

define the period matrix which is an element of the Siegel upper half-plane $\mathbb{H}_g$, that is symmetric with imaginary part positive definite. The column of Ω define an integer lattice T_Ω and an associated Jacobian variety $J_\Omega = \mathbb{C}^g/T_\Omega$. Conversely, from Ω one can recover the starting Riemann surface. Inserting into the algebraic geometric expression (??) one finds:

$$Z^g_{BOS} = c_g \int | \prod_{I \leq J} d\Omega_{IJ}|^2 \det(Im\Omega)^{-13} |\Phi(\Omega)|^{-2},$$

$\Phi = 1/F$. The partition function Z^g_{BOS} is independent from the choice of the canonical basis[3]. Two different canonical basis differ by a symplectic transformation

$$\begin{pmatrix} A'_I \\ B'_J \end{pmatrix} = M \begin{pmatrix} A_I \\ B_J \end{pmatrix},$$

$$M = \begin{pmatrix} U & T \\ V & Z \end{pmatrix} \in Sp(2g, \mathbb{Z}).$$

Correspondingly the period matrix transform as

$$\Omega \longrightarrow (U\Omega + T)(V\Omega + Z)^{-1}.$$

It then follows that necessarily Φ is a modular form of weight $12 - g$

$$\Phi(\Omega) \longrightarrow \Phi(\Omega)(\det(V\Omega + Z))^{12-g}.$$

For example, at genus $g = 2$, $\Phi(\Omega)$ has weight ten. A theorem due to Igusa states that at genus two, modular forms realize a polynomial ring with four generators ψ_i, of weights $i = 4, 6, 10, 12$. Thus we must have

$$\Phi(\Omega) = \alpha\psi_4\psi_6 + \beta\psi_{10}.$$

On the other hand, the partition function must satisfy the clustering condition, that is if we pinch the surface Σ_2 separating it into the union of two tori $\Sigma_{g=2} \to \Sigma_1(\Omega_1) \cup \Sigma_1(\Omega_2)$, with complex parameters Ω_1 and Ω_2, then the partition function factorizes as $Z_2 = Z_1 Z_1$. This selects $\alpha = 0$ so that

$$\Phi(\Omega) = \psi_{10} = \prod_{\delta \text{ even}} \theta[\delta](0, \Omega)$$

[3]The local expression for the global section depend on such a choice, but not the global section itself.

and

$$Z^g_{BOS} = c_2 \int |\prod_{I\leq J} d\Omega_{IJ}|^2 \prod_{\delta \text{ even}} |\theta[\delta](0,\Omega)|^{-2} (\det Im\Omega)^{-13}.$$

Here $\theta[\delta](z,\Omega)$, $z \in \mathbb{C}^g$ (the covering space of the Jacobian variety J), $\Omega \in \mathbb{H}_g$, are the theta function with characteristic[4] $[\delta] = [^a_b] \equiv a + \Omega b$, $a, b \in \mathbb{Z}^g$ defined by

$$\theta[\delta](z,\Omega) = e^{i\pi b\cdot(z+\frac{a}{2})+i\frac{\pi}{4}b\cdot\Omega b}\theta(z + a + \Omega b, \Omega),$$
$$\theta(z,\Omega) = \sum_{m\in\mathbb{Z}^g} e^{2i\pi m\cdot z + i\pi m\cdot\Omega m}.$$

Conclusion

What we have learned from the bosonic history is that, passing through the algebraic-geometric description, one is led to a global description of the amplitudes measure. This provides a rigorous and almost well defined expression, the only impediment being represented by the divergence due to the pole at infinity of the section. This can be indeed imputed to the presence of the tachyon in the bosonic spectrum.
By means of the GSO projection the tachyon disappears from the spectrum of supersymmetric strings so that we could expect that a similar treatment extended to the supersymmetric case should lead to a completely well defined expression for the partition function. Unfortunately supersymmetry makes all things much more difficult and actually an analogue global description is not yet available. We will see now why this happens and a possible strategy proposed by D'Hoker and Phong and worked out at genus two [DP1], [DP2], [DP3], [DP4].

1.2. Supersymmetric Strings

We can try to generalize directly the bosonic realization to the supersymmetric case. On the Riemann surface Σ_g there are 2^{2g} possible spin structures. For any choice among them, one can define spinor fields over the surface. To the metric h_{ab} one can then define its superpartner χ^α_a, the gravitino. The coordinate fields x^μ have as superpartners Majorana spinors ψ^μ. For any fixed spin structure one can then define the superstring action

$$\begin{aligned} I_{g,\delta} &= \frac{1}{4\pi\alpha'}\int_{\Sigma_g} d^2z\sqrt{h}[\frac{1}{2}h^{\alpha\beta}\partial_\alpha x^\mu \partial_\beta x_\mu - \frac{i}{2}\psi^\mu\gamma^\alpha D_\alpha\psi_\mu \\ &- \frac{1}{2}\psi^\mu\gamma^a\gamma^\alpha\chi_a\partial_\alpha x_\mu + \frac{1}{8}\psi^\mu\gamma^a\gamma^b\chi_a(\chi_b\psi_\mu)] + \lambda\mathcal{X}(\Sigma_g), \end{aligned}$$

where $\mathcal{X}$ is the Euler characteristic. Here we reserved Greek indices for tangent directions and Latin indices for flat direction. In particular the metric is related to a zweibein e^a_α via the usual relation $g_{\alpha\beta} = e^a_\alpha e^b_\beta \delta_{ab}$. The gamma matrices satisfy $\{\gamma^a, \gamma^b\} = -2\delta^{ab}$. The covariant derivative is $D_\alpha\psi_\mu = \partial_\alpha - \frac{i}{2}\omega_\alpha\gamma^1\gamma^2$, ω being the spin connection. The spinor fields behave as half-integer powers of differentials so that we can write

$$\psi = \psi_+ dz^{\frac{1}{2}}, \qquad \bar{\psi} = \psi_- d\bar{z}^{\frac{1}{2}}, \qquad \chi = \chi^+_{\bar{z}} d\bar{z} dz^{-\frac{1}{2}}, \qquad \bar{\chi} = \chi^-_z dz d\bar{z}^{-\frac{1}{2}}.$$

[4] In the second member a, b ar thought as row vectors and in the last as column vectors.

The symmetries of the action are thus extended to supersymmetries by adding to diffeomorphisms $\mathrm{Diff}(\Sigma_g)$

$$\begin{aligned} \delta_v e^a_\alpha &= v^\beta \partial_\beta e^a_\alpha + e^a_\beta \partial_\alpha v^\beta, \\ \delta_v \chi_a &= v^\beta \partial_\beta \chi_a + \chi_b e^\alpha_a \partial_\alpha v^b, \\ \delta_v x^\mu &= v^\alpha \partial_\alpha x^\mu, \\ \delta_v \psi^\mu &= v^\alpha \partial_\alpha \psi^\mu, \end{aligned}$$

and Weyl transformations $\mathrm{Weyl}(\Sigma_g)$

$$\begin{aligned} \delta_\omega e^a_\alpha &= \omega e^a_\alpha, \\ \delta_\omega \chi_a &= \frac{\omega}{2}\chi_a, \\ \delta_\omega x^\mu &= 0, \\ \delta_\omega \psi^\mu &= -\frac{\omega}{2}\psi^\mu, \end{aligned}$$

the supersymmetry transformations $\mathrm{Susy}(\Sigma_g)$

$$\begin{aligned} \delta_\epsilon e^a_\alpha &= i\epsilon\gamma^a\chi_\alpha, \\ \delta_\epsilon \chi_a &= 2D_a\epsilon, \\ \delta_\epsilon x^\mu &= \epsilon\psi^\mu, \\ \delta_\epsilon \psi^\mu &= -\frac{i}{2}(\gamma^\beta\epsilon)(\partial_\beta x^\mu - \frac{1}{2}\chi_\beta\psi^\mu), \end{aligned}$$

and super-Weyl transformations $\mathrm{SW}(\Sigma_g)$

$$\begin{aligned} \delta_\lambda e^a_\alpha &= 0, \\ \delta_\lambda \chi_a &= \gamma_a\lambda, \\ \delta_\lambda x^\mu &= 0, \\ \delta_\lambda \psi^\mu &= 0. \end{aligned}$$

Here ϵ and λ are spinors and all spinor index, which we have omitted everywhere, are contracted in an obvious way along the $NW - SE$ convention.
One has thus to compute the partition function (for fixed spin structure δ)

$$Z^g_\delta = \int [Dh_{\alpha\beta}][D\chi_a][Dx^\mu][D\psi^\mu] \exp(-I_{g,\delta}). \tag{6}$$

The measure is the one inherited by the bosonic metric plus the metric for spinor deformations

$$\|\delta\psi\|^2 = \int_{\Sigma_g} \delta\bar\psi^\mu \delta\psi_\mu \sqrt{h} d^2\zeta, \qquad \|\delta\chi_a\|^2 = \int_{\Sigma_g} \delta\bar\chi_a \delta\chi_b h^{ab} \sqrt{h} d^2\zeta.$$

Moreover, it is convenient to introduce a norm in the space of supersymmetry deformations

$$\|\epsilon\|^2 = \int_{\Sigma_g} \bar\epsilon\epsilon\sqrt{h} d^2\zeta.$$

Again, we can use the huge symmetry group to reduce the path integral to a finite dimensional integration. The moduli space to be considered is now

$$\mathcal{SM}_g = (\{h_{\alpha\beta}\} \times \{\chi_a\})/(\mathrm{Diff}(\Sigma_g) \times \mathrm{Weyl}(\Sigma_g) \times \mathrm{Susy}(\Sigma_g) \times \mathrm{SW}(\Sigma_g)). \tag{7}$$

To better understand it one can study its tangent bundle (the super Teichm'üller space). Locally it splits into the usual bosonic Teichm'üller space plus a novel part described by the gravitinos deformations. Under symmetry transformations gravitinos deform as

$$\delta_{(v,\omega,\epsilon,\lambda)}\chi_a = \frac{\omega}{2}\chi_a + \gamma_a\lambda + v^\beta\partial_\beta\chi_a + \chi_\beta e_a^\alpha\partial_\alpha v^\beta + 2D_a\epsilon.$$

Genuine deformations $\delta_\perp\chi_a$ are then defined by the spinor deformations orthogonal to symmetry deformations. Let us define the operator $P_{\frac{1}{2}}$ sending $\frac{1}{2}$ spinors to $\frac{3}{2}$ spinors, defined by

$$[P_{\frac{1}{2}}\epsilon]_a := 2D_a\epsilon + \gamma_a\gamma^\beta D_\beta\epsilon.$$

Its adjoint $P_{\frac{1}{2}}^\dagger$, acting on $\frac{3}{2}$ spinors, is defined by

$$\langle P_{\frac{1}{2}}^\dagger\delta\chi, \epsilon\rangle = \langle\delta\chi, P_{\frac{1}{2}}\epsilon\rangle,$$

where the scalar products are the ones associated to the metrics defined above. It results that $\{\delta_\perp\chi\} = \mathrm{Ker}P_{\frac{1}{2}}^\dagger$. Moreover $\mathrm{Ker}P_{\frac{1}{2}}^\dagger = V^{\frac{3}{2}}$, the holomorphic $\frac{3}{2}$-differentials. Indeed, it happens that $V^2 \oplus V^{\frac{3}{2}}$ can be seen as a superspace, so that the true deformations of gravitinos are measured by the superpartners of the quadratic differentials. An application of the Riemann-Roch theorem shows that the super moduli space is a complex super manifold of dimension[5] $(3g-3|2g-2)$. We will not investigate this further, all details can be found in the lecture notes of D'Hoker and Phong [DP7], see also Nelson [GN1, GN2] for a deeper analysis of super moduli spaces. We only mention that, correspondingly, the Beltrami differential will be extended to super-Beltrami differentials whose odd components $\Theta_{a,u}$, similarly to the even ones, will parameterize the gravitino deformations so that, in local coordinates $(t^i, \theta^u) \in \mathbb{C}^{(3g-3|2g-2)}$,

$$\Theta_{a,u} = \frac{\partial\chi_a}{\partial\theta^u} + \frac{1}{2}\gamma_a\gamma^b\frac{\partial\chi_b}{\partial\theta^u},$$

and the orthogonal deformations are

$$\delta_\perp\chi_a = \Gamma_{vw}\langle\Pi^w, \Theta_u\rangle\Pi_a^v\theta^u.$$

Here $\{\Pi_a^u\}_{u=1}^{2g-2}$ is a basis of $V^{\frac{3}{2}}$, and Γ_{vw} is the inverse matrix of $\Gamma^{vw} = \langle\Pi^v, \Pi^w\rangle$. Proceeding as for the bosonic case one finally arrive to the expression

$$\begin{aligned} Z_\delta^g &= \int [Dx^\mu][D\psi^\mu]d\mu_t d\mu_\theta \det\Delta_{-1}\frac{|\det\langle\Phi^i,\mu^j\rangle|^2}{\det\langle\Phi^i,\Phi^j\rangle}\cdot \\ &\quad \cdot\det\Delta_{-\frac{1}{2}}\frac{|\det\langle\Pi^u,\Theta^v\rangle|^2}{\det\langle\Pi^u,\Pi^v\rangle}e^{-I_{g,\delta}}, \end{aligned} \tag{8}$$

[5] We mention here only the case of genus $g \geq 2$.

where $d\mu_t = dt^{3g-3}d\bar{t}^{3g-3}$ and $d\mu_\theta = d\theta^{2g-2}d\bar{\theta}^{2g-2}$. Note that we have yet integrated out the symmetry group degrees of freedom. Again, this can be verified computing the Weyl variation. One obtains that in this case anomalies disappear if $D = 10$. Using the symmetry transformations, we can project the gravitinos on the $V^{\frac{3}{2}}$ part so that

$$\chi_a = \sum_{u=1}^{2g-2} \zeta_u \Pi_a^u,$$

where ζ_u are fermionic coordinates, which we will use in place of the θ^u (this give not any nontrivial contribution to the Jacobian). In particular this imply $\gamma^a\chi_a = 0$ so that the action takes the simpler form[6]

$$I_{g,\delta} = \int_{\Sigma_g} d^2z\sqrt{h}[\frac{1}{2}h^{\alpha\beta}\partial_\alpha x^\mu \partial_\beta x_\mu - \frac{i}{2}\psi^\mu\gamma^\alpha D_\alpha\psi_\mu + \psi^\mu\chi^\alpha\partial_\alpha x_\mu$$
$$-\frac{1}{4}\psi^\mu\chi^a\psi_\mu\chi_a].$$

We can now first integrate over the x^μ configurations. This gives

$$\int[Dx^\mu]e^{-I_\delta} = \left(\frac{\det\Delta_0}{\int h_{z\bar{z}}d^2\zeta}\right)^{-\frac{D}{2}} e^{\frac{i}{2}\int\sqrt{h}\psi^\mu\gamma^\alpha D_\alpha\psi_\mu d^2z} e^{-\frac{1}{2}\sum_{u,v}\zeta_u\zeta_v K^{uv}},$$

$$K^{uv} = \int d^2zd^2z'\sqrt{h(z,\bar{z})h(z',\bar{z}')}\psi^\mu\Pi^{u,a}(z)\psi_\mu\Pi^{v,b}(z')\Sigma_{a,b}(z,z'),$$

$$\Sigma_{a,b}(z,z') = \frac{\partial^2}{\partial z^a\partial z'^b}\Delta_0^{-1} - \frac{1}{2}h_{ab}\delta(z,z').$$

Integration over ζ_u then gives the Pfaffian of K, and, finally, integration over ψ^μ gives

$$Z_\delta^g = \int d\mu_t \det\Delta_{-1}\frac{|\det\langle\Phi^i,\mu^j\rangle|^2}{\det\langle\Phi^i,\Phi^j\rangle}\det\Delta_{-\frac{1}{2}}\frac{|\det\langle\Pi^u,\Theta^v\rangle|^2}{\det\langle\Pi^u,\Pi^v\rangle}$$
$$\cdot\left(\frac{\det\Delta_0}{\int h_{z\bar{z}}d^2\zeta}\right)^{-5}(\det{}'(\gamma^a D_a))^5\int[D\psi_0^\mu] < \mathrm{Pfaff}K >_{\psi'}, \tag{9}$$

where a prime indicates that zero modes are dropped, ψ_0^μ are zero modes of $\gamma^a D_a$ and $< \mathrm{Pfaff}K >_{\psi'}$ indicates expectation value.

1.2.1. An Unfortunate History

Expression (**??**) is conformally invariant and apparently gives the desired result. However, here is where the problems start. A first problem is to realize a chiral splitting. This is a first point where the absence of holomorphic anomaly should be very helpful. Indeed, the obtained expression is only for fixed spin structure δ. To define the full super string theory model one need to sum up over all spin structures taking account of the GSO projection in order to eliminate the tachyon. But GSO projection acts separately on the chiral modes so

[6]We also set $4\pi\alpha' = 1$ for simplicity, and omitted the topological term.

that the splitting becomes essential.
But even before to solve this problem, one must recognize that (??) is ambiguous.
One expects for the partition function to be independent from the choice of the parametrization of the moduli, that is from the choice of the super Beltrami differentials. Any change should eventually add to the integral some boundary terms which should vanish. But this is not what happen: the boundary terms do not generically vanish.
The ambiguity was first noted by Verlinde and Verlinde [VV]. In [MM], Moore and Morozov analyzed the problem on the light of some consistency conditions superstring theories should satisfy: modular invariance, vanishing of the cosmological constant, and nonrenormalization theorems. In particular, they have computed the difference between type II and heterotic partition functions for genus two surfaces, showing that they differ by a positive term, so that they seem to be not simultaneously consistent. The physical origin of the ambiguity has been further investigated in [ARS], se also Morozov and Perelomov [MP]. Here they computed the $g = 2$ heterotic partition function by choosing an explicit basis for the super Beltrami differentials, represented by δ-functions supported on fixed points z_u $(u = 1, 2)$. It then results an explicit dependence on the points in the sense that changing the points gives rise to a shift of the integrand by a differential term which does not vanish on the boundary. The boundary of the moduli space contains two disjoint tori with two marked points p_u. There, it is shown that the ambiguity disappears if one choose z_u in such the way that $z_u = p_u$ on the boundary.
A decisive analysis of the ambiguities can be found in [AMS], where global issues are considered. They found that the superstring measure is a total derivative so that all problems are related to the boundary conditions. Many peculiarities of the ambiguities are put in light but the analysis do not provide a prescription able to eliminate them. We will demand to the cited literature all details and will not discuss it here further.

1.2.2. The D'Hoker and Phong Strategy

A proposal for solving the ambiguity problem comes from D'Hoker and Phong [DP7]. The main idea is that the problems have origin in a wrong choice for the slice parametrization, that is the choice of the metric to select the bosonic component of the slice is not a good one. Suppose that a slice is selected by a choice $(h_{\alpha\beta}, \chi_\alpha)$. After a supersymmetric transformation one obtains a new coordinatization $(\tilde{h}_{\alpha\beta}, \tilde{\chi}_\alpha)$. If the metric were a good selection for the bosonic components, then the projector $\phi : (h_{\alpha\beta}, \chi_\alpha) \mapsto h_{\alpha\beta}$ would be supersymmetry preserving in the sense that the supersymmetry transformation would induce a diffeomorphism (eventually composed with a conformal transformation) $h_{\alpha\beta} \mapsto \tilde{h}_{\alpha\beta}$. But this results not to be the case and in general $h_{\alpha\beta}$ and $\tilde{h}_{\alpha\beta}$ are not related by a bosonic symmetry.
Their main idea is to substitute the metric with the period matrix associated to the Riemann surface to be considered. As we said, the choice of a canonical basis for holomorphic differentials, associated to a given symplectic basis $\{A_I, B_J\}$ of $H_1(\Sigma_g, \mathbb{Z})$ defines a $g \times g$ period matrix $\Omega = \{\Omega_{IJ}\} \in M_g(\mathbb{C})$. It can be shown that the period matrix lies in the Siegel upperhalf space

$$\Omega \in \mathbb{H}_g := \{\tau \in M_g(\mathbb{C}) : \tau = {}^t\tau,\ \mathrm{Im}\tau > 0\}. \tag{10}$$

The important fact is that Torelli's theorem stating that Σ_g is completely characterized by its period matrix.
After introducing a suitable superdifferential description of super-Riemann surfaces, D'Hoker and Phong introduce the concept of super holomorphic differentials $\hat{\omega}_I = \omega_{I0} + \theta\omega_{I1}$, θ being an odd variable, which are associated to a symplectic basis by

$$\oint_{A_I} dz d\theta \hat{\omega}_J = \delta_{IJ},$$

and define the super period matrix

$$\hat{\Omega}_{IJ} = \oint_{B_I} dz d\theta \hat{\omega}_J.$$

Indeed, they showed that

$$\hat{\Omega}_{IJ} = \Omega_{IJ} - \frac{i}{8\pi}\int d^z d^2 z' \omega_{I0}(z)\chi_{\bar{z}}^+ S_\delta(z,z')\chi_{\bar{z}'}^+ \omega_{J0}(z'),$$

where S_δ is the Szegö kernel, the unique solution of the equation

$$\partial_{\bar{z}} S_\delta(z,z') + \frac{1}{8\pi}\chi_{\bar{z}}^+ \int d^2 w \chi_{\bar{w}}^+ \partial_z \partial_w \ln E(z,w) S_\delta(w,z') = 2\pi\delta(z,z'),$$

and $E(z,w)$ is the prime form. $\hat{\Omega}$ is indeed supersymmetric and does not suffer the defects of the metric. In a long series of remarkable papers, they have been able to prove that the super period matrix prescription provides a well defined result for the amplitudes. Most of calculation are explicitly developed for the genus two case, but they argued that in principle they should work at any g. For $g = 2$, in particular, slice independence has been verified as well as the nonrenormalization theorems. After integrating over the odd moduli, a nice expression for the $g = 2$ whole vacuum amplitude was found to be

$$Z^2 = \int_{\mathcal{M}_2} (\det \operatorname{Im} \Omega)^{-5} \sum_{\delta\delta'} c_{\delta\delta'} d\mu[\delta](\Omega) \wedge \overline{d\mu[\delta'](\Omega)}, \tag{11}$$

where $c_{\delta\delta'}$ are phases realizing the right GSO projection and

$$d\mu[\delta](\Omega) = \frac{\theta^4[\delta](0,\Omega)\Xi_6[\delta](\Omega)}{16\pi^6\psi_{10}(\Omega)} \prod_{I\leq J} d\Omega_{IJ}, \tag{12}$$

$$\Xi_6[\delta](\Omega) := \sum_{1\leq i<j\leq 3} \langle \nu_i|\nu_j\rangle \prod_{k=4,5,6} \theta^4[\nu_i+\nu_j+\nu_k](0,\Omega)\,, \tag{13}$$

where each even spin structure[7] δ is written as a sum of three distinct odd spin structures $\delta = \nu_1 + \nu_2 + \nu_3$ and ν_4, ν_5, ν_6 denote the remaining three distinct odd spin structures, and

$$\langle\kappa|\lambda\rangle := e^{\pi i(a_\kappa\cdot b_\lambda - b_\kappa\cdot a_\lambda)}, \qquad \kappa = \left[\begin{smallmatrix} a_\kappa \\ b_\kappa \end{smallmatrix}\right], \lambda = \left[\begin{smallmatrix} a_\lambda \\ b_\lambda \end{smallmatrix}\right].$$

[7]In what follow we will indicate spin structures of any genus with Δ, just for the case $g = 2$ we will use, as D'Hoker and Phong, δ.

Finally, ψ_{10} is the Igusa form

$$\psi_{10} = \prod_{\delta \text{ even}} \theta[\delta](0,\Omega). \tag{14}$$

D'Hoker and Phong claimed that after integrating over odd moduli:

$$Z^g = \int_{\mathcal{M}_g} (\det Im\Omega)^{-5} \sum_{\Delta\Delta'} c_{\Delta\Delta'} d\mu[\Delta](\Omega) \wedge \overline{d\mu[\Delta'](\Omega)} \tag{15}$$

with

$$d\mu[\Delta](\Omega) = d\mu_{BOS}(\Omega)\Xi_8(\Delta),$$
$$\Delta = \begin{bmatrix} a \\ b \end{bmatrix}, \qquad a, b \in \mathbb{Z}_2^g,$$

and $\Xi_8[\Delta](\Omega)$ are equivariant modular forms[8] (the symplectic group acts on the characteristics Δ also). In [DP5] a detailed analysis of the $g = 2$ results led D'Hoker and Phong to make an ansatz for the measure at genus 3. The genus three bosonic measure is

$$\mathrm{d}\mu_B^{(3)} = \frac{c_3}{\Psi_9(\Omega)} \prod_{I \leq J} \mathrm{d}\Omega_{IJ}$$

where $\Psi_9^2(\Omega)$ is a Siegel modular form of weight 18 for $Sp(6,\mathbb{Z})$. By analogy with the $g = 2$ case, D'Hoker and Phong [DP5] proposed that the genus three chiral superstring measure is of the form

$$\mathrm{d}\mu[\Delta] = \frac{\theta[\Delta]^4(0,\Omega)\Xi_6[\Delta](\Omega)}{8\pi^4\Psi_9(\Omega)} \prod_{I \leq J} d\Omega_{IJ},$$

and they gave three constraints on the functions $\Xi_6[\Delta^{(3)}](\tau^{(3)})$. Note that they assume $\Xi_8[\Delta](\Omega) \equiv \theta[\Delta]^4(0,\Omega)\Xi_6[\Delta](\Omega)$. However, they did not succeed in finding functions which satisfy all their constraints [DP6].

The failure in finding such a solution is not simply due to the formidable problem of searching it in a very big space of modular forms without making use of a systematic procedure, but mainly on the fact that it does not exist, as we will prove in these notes. In particular, as was also remarked by Morozov, the main obstacle in finding a solution was the too strong and prejudicious imposition for the measure to be proportional to the fourth power of $\theta[\Delta](0,\Omega)$. After eliminating this condition and adapting the D'Hoker and Phong constraints to the most general ansatz, led us to determine the existence of a unique solution at genus 3 and at genus 4. This will be the main argument of the rest of these notes.

The notes are organized as follows. In section 2. we present the general ansatz for the measure also providing some criticisms. In section 3. are given the mathematical notions and instruments necessary to tackle the problem of finding a solution of the ansatz. In section 4. we use the developed mathematical tools to reformulate the ansatz in a simpler form. Finally, in section 5. we determine the solution for $g \leq 4$, and prove its strict uniqueness for $g < 4$. For $g = 4$ the uniqueness is given in a weaker form. Some further comments are presented in Conclusion.

[8]Will be much more detailed in the next sections.

2. The General Ansatz

Our starting point consists in assuming the validity of (11) to be true. This is a crucial point so that some criticisms have been to be considered. Before discussing such points, let us finish to expose our approach. As discussed by Morozov [Mo] there are two strategies to deal with superstring measures. The first one is the more direct one, that is by direct integration of odd moduli after holomorphic factorization. The second one is to use general considerations to deduce a reasonable guess for the measure and then to use its properties to determine its final form. We will follow this second approach.
We will assume the problem of computing the bosonic measure as resolved and well known. We have seen that the bosonic partition function density is

$$|d\mu^{(g)}_{BOS}|^2(\det \mathrm{Im}\Omega)^{-13} = |\prod_{I\leq J} d\Omega_{IJ}|^2 \det(\mathrm{Im}\Omega)^{-13}|\Phi(\Omega)|^{-2}.$$

Invariance under modular transformations requires precise modular properties for $d\mu_{BOS}$. This structure is very strongly supported by global issues in algebraic and complex geometry. For super symmetric strings one instead obtains the much more less supported expression (11). If the Belavin-Knizhnik and Manin-Mumford arguments really play a crucial role in superstring theory too, then, looking again at modular properties, it is natural to expect for the superstring measure at fixed chirality to be related to the bosonic measure (Mumford form) by the relation

$$d\mu^{(g)}[\Delta^{(g)}] = \Xi_8^{(g)}[\Delta^{(g)}](\Omega)d\mu^{(g)}_{BOS}, \tag{16}$$

so that all dependence on the spin structure, that is the characteristic Δ, is codified in the factor $\Xi_8[\Delta](\Omega)$. Let as look at the known examples. We specify the genus by an apex (g), for example $\Delta^{(1)}$ means a $g = 1$ characteristic. At genus one the chiral measure is

$$d\mu[\Delta^{(1)}] = c^{(1)}\theta[\Delta^{(1)}]^4(\Omega^{(1)})\eta^{12}(\Omega^{(1)})\, d\mu^{(1)}_B, \tag{17}$$

$$d\mu^{(1)}_{BOS} = \frac{1}{(2\pi)^{12}\eta^{24}(\Omega^{(1)})}d\Omega^{(1)}. \tag{18}$$

Then $\Xi_8[\Delta^{(1)}](\Omega^{(1)}) = \theta[\Delta^{(1)}]^4(\Omega^{(1)})\eta^{12}(\Omega^{(1)})$ is a modular form of weight 8 on a certain subgroup of $SL(2,\mathbb{Z})$ and $\eta(\Omega)$ is the Dedekind function, see section 5.1..
For genus two D'Hoker and Phong obtained

$$d\mu[\Delta^{(2)}] = c^{(2)}\theta[\Delta^{(2)}]^4(\Omega^{(2)})\Xi_6[\Omega^{(2)}](\Omega^{(2)})\, d\mu^{(2)}_{BOS}, \tag{19}$$

$$d\mu^{(2)}_{BOS} = \frac{c_2}{\Psi_{10}(\Omega^{(2)})}\prod_{i\leq j} d\Omega_{ij}, \tag{20}$$

where Ξ_6 has been defined in the previous section and

$$\Xi_8[\Delta^{(1)}](\Omega^{(1)}) = \theta[\Delta^{(2)}]^4(\Omega^{(2)})\Xi_6[\Delta^{(2)}](\Omega^{(2)}) \tag{21}$$

is indeed a modular form of weight 8 on a suitable subgroup of $Sp(4,\mathbb{Z})$. It also has the right behavior at the boundary of the moduli space. For example, in the limit where the genus two

Riemann surface Σ_2 splits as the union of two elliptic curves $\Sigma_1[\Omega_1^{(1)}]$ and $\Sigma_1[\Omega_2^{(1)}]$ with moduli given by $\Omega_1^{(1)} = \Omega_{11}^{(2)}$ and $\Omega_2^{(1)} = \Omega_{22}^{(2)}$, then the measure separates as the product of the genus one measures. Such limiting behavior, which we will call the clustering property, is a fundamental property to be satisfied by the right superstring measures.
Following the declared strategy we are now ready to state a general guess for the supersymmetric invariant measure at any genera, t.i. for the functions $\Xi_8[\Delta^{(g)}](\Omega^{(g)})$. This consists in three points:
(i) The functions $\Xi_8[\Delta^{(g)}]$ are holomorphic on the Siegel upper halfplane $\mathbb{H}_g$.
(ii) Under the action of the symplectic group $Sp(2g,\mathbb{Z})$ on $\mathbb{H}_g$, they should transform as follows:

$$\Xi_8[M\cdot\Delta^{(g)}](M\cdot\Omega) = \det(C\Omega+D)^8\Xi_8[\Delta^{(g)}](\Omega), \tag{22}$$

for all $M\in Sp(2g,\mathbb{Z})$. Here the affine action of M on the characteristic $\Delta^{(g)}$ is given by

$$\begin{pmatrix} A & B \\ C & D \end{pmatrix} \cdot \begin{bmatrix} a \\ b \end{bmatrix} := \begin{bmatrix} c \\ d \end{bmatrix}, \tag{23}$$

$$\begin{pmatrix} {}^t c \\ {}^t d \end{pmatrix} = \begin{pmatrix} D & -C \\ -B & A \end{pmatrix} \begin{pmatrix} {}^t a \\ {}^t b \end{pmatrix} + \begin{pmatrix} (C\,{}^tD)_0 \\ (A\,{}^tB)_0 \end{pmatrix} \mod 2$$

where $N_0 = (N_{11},\dots,N_{gg})$ is the diagonal of the matrix N.
(iii) The restriction of these functions to 'reducible' period matrices is a product of the corresponding functions in lower genus. More precisely, let

$$D_{k,g-k} := \left\{ \Omega_{k,g-k} := \begin{pmatrix} \Omega_k & 0 \\ 0 & \Omega_{g-k} \end{pmatrix} \in \mathbb{H}_g \ : \ \Omega_k \in \mathbb{H}_k,\ \Omega_{g-k}\in\mathbb{H}_{g-k} \right\}$$
$$\cong \mathbb{H}_k \times \mathbb{H}_{g-k}.$$

Then we require that for all k, $0<k<g$,

$$\Xi_8\begin{bmatrix} a_1\dots a_k\ a_{k+1}\dots a_g \\ b_1\dots b_k\ b_{k+1}\dots b_g \end{bmatrix}(\Omega_{k,g-k}) = \Xi_8\begin{bmatrix} a_1\dots a_k \\ b_1\dots b_k \end{bmatrix}(\Omega_k)\Xi_8\begin{bmatrix} a_{k+1}\dots a_g \\ b_{k+1}\dots b_g \end{bmatrix}(\Omega_{g-k})$$

for all even characteristics $\Delta^{(g)} = \begin{bmatrix} a_1\dots a_g \\ b_1\dots b_g \end{bmatrix}$ and all $\Omega_{k,g-k}\in D_{k,g-k}$.

Note that our guess coincide with the one of D'Hoker and Phong apart from the fact that we do not require the functions $\Xi_8[M\cdot\Delta^{(g)}]$ to factorize as the products of $\theta[\Delta^{(3)}]^4(\Omega^{(3)})$ times some equivariant modular form of weight 6. This is exactly what will permit us to make success in finding a (unique) solution at genus 3 and 4.
A direct tackling of this constraints is quite formidable and can explain the failure of D'Hoker and Phong in finding a solution (or recognizing that their ansatz was to strong). For this reasons we will take advantage of the theory of induced representations: we will build up representations of the modular group on the space of forms starting from the representations given by a suitable subspace left invariant by a certain subgroup of the entire modular group. This way to proceed resemble the method used by Wigner to classify the irreducible representations of the Poincaré group induced from the representations of the little group.
It is clear that this program requires an adequate exposition of the mathematical tools necessary to reach the target. We will provide all the necessary mathematical background in

the next section. Before to do it, let us stop for a moment to expose the promised criticisms to formula (11).

To obtain the result (11) one has to integrate over the odd coordinates of the moduli space. This is the super moduli space of super Riemann surfaces. To this aim one needs a "splitting" of the super-Riemann surface $\hat{\Sigma}$. For example, at $g = 2$ and for even spin structures this should be done as follows:

- find a basis for super Abelian differentials ω_1, ω_2, with only even part;
- take periods $\hat{\Omega}$ and the Jacobian variety $J = \mathbb{C}^2/\hat{\Omega}$;
- take for Σ the Riemann surface having J as Jacobian;

then $\hat{\Sigma} \longrightarrow \Sigma$ is a fibration and the point in $\mathcal{M}_2$ define the splitting of the super moduli space. For $g > 2$ the situation is quite complicated, and it is hard to argue that a similar splitting should work in this case.[9] Generically, it happens that odd differentials do not exist. However, some odd differential may exist for special complex structures. In this case the Jacobian variety is no more well defined and this procedure breaks down. But let us assume that there were not odd differentials, so that $J = \mathbb{C}^g/\hat{\Omega}$ is well defined. But there is not any particular reason to believe for it to be the Jacobian variety of ordinary Riemann surface Σ_g: its periods can differ from those of an ordinary Riemann surface by terms that are bilinear in fermionic moduli.
To make this more clear, let us recall that the ordinary period matrix Ω is a point in $\mathbb{H}_g$. On this space it acts the modular group $\Gamma_g := Sp(2g, \mathbb{Z})$. Following [vG1], let us define $A_g := \Gamma_g \backslash \mathbb{H}_g$. Torelli's theorem than ensures that the natural holomorphic map $j : \mathcal{M}_g \to A_g$ is injective. If $\mathbb{J}_g^0 \subset \mathbb{H}_g$ is the set of all period matrices of genus g Riemann surfaces, the Jacobian locus is its closure $\mathbb{J}_g$ in $\mathbb{H}_g$. It can be shown that $\mathbb{J}_g - \mathbb{J}_g^0 \subset \mathbb{H}_g$ consists of block diagonal matrix whose diagonal blocks are period matrices of lower dimensional Riemann surfaces. As we have seen, in string theory these select the degenerate limits which play an important role in computing the amplitudes (clustering property). Therefore, we are really interested in considering the Jacobi locus $\mathbb{J}_g$ or better its image j_g in A_g. It results that $\mathbb{J}_g$ is a $(3g-3)$-dimensional complex subvariety of $\mathbb{H}_g$. The Schottky problem is related to the question of what points of $\mathbb{H}_g$ are in $\mathbb{J}_g$. In Table 1 we report the dimensions of these manifolds for inreasing g. From this it follows that the Schottky problem is trivial up to genus three. To ensure the validity of (11) at genus 3 we then should worry only about the existence of odd differentials. To higher genera the situation complicates because the codimension of $\mathbb{J}_g$ in $\mathbb{H}_g$ increases quadratically with g, so that we cannot expect, without some strong motivation, for the super period matrix $\hat{\Omega}$ to lie in $\mathbb{J}_g$. Thus, even though we will see that the ansatz provide a solution for the $g = 3, 4$ cases, a much deeper investigation must be devoted to understand (11) or improve.

[9]This observations was pointed out to us by Ed. Witten to which we are very grateful for his explanations.

Table 1. Dimensions of $\mathbb{H}_g$ and $\mathcal{M}_g$.

g	$\dim\mathbb{H}_g$	$\dim\mathcal{M}_g$
0	0	0
1	1	1
2	3	3
3	6	6
4	10	9
g	$g(g+1)/2$	$3g-3$

3. Mathematical Background

In order to describe the construction of the string amplitudes from an axiomatic point of view we need to develop some mathematical tools. In this section we will introduce the symplectic group and modular forms, the theta functions and their transformation properties under the action of the symplectic group.

3.1. The Symplectic Group

The symplectic group $\mathrm{Sp}(2g, F)$ of degree $2g$ over a field F is the group of $2g \times 2g$ matrices with entries in F satisfying:

$$ME\,{}^tM = E, \tag{24}$$

with $E = \begin{pmatrix} 0 & I_g \\ -I_g & 0 \end{pmatrix}$ the canonical symplectic form and I_g the g-dimensional identity matrix. Note that the symplectic form has determinant $+1$, its inverse is $E^{-1} = {}^tE = -E$ and that a symplectic matrix M is always invertible and its inverse is $M^{-1} = E^{-1}\,{}^tME$. Also the product of two symplectic matrices is symplectic: suppose $M = M_1M_2$, with M_1 and M_2 symplectic thus $ME\,{}^tM = M_1M_2E\,{}^tM_2\,{}^tM_1 = M_1E\,{}^tM_1 = E$. This shows that $\mathrm{Sp}(2g, F)$ with matrix multiplication is a group. Directly from the definition, it follows that the determinant of a symplectic matrix is ± 1, but it turns out that this determinant is always positive. To see this one uses the identity [10] $\mathrm{Pf}({}^tMEM) = \det(M)\,\mathrm{Pf}(E)$, since ${}^tMEM = E$ and $\mathrm{Pf}(E) \neq 0$ it follows that $\det(M) = 1$. Thus, the symplectic group is a subgroup of the special linear group $\mathrm{SL}(2g, F)$. For a block matrix $M = \begin{pmatrix} A & B \\ C & D \end{pmatrix}$ the condition to be symplectic is equivalent to the conditions

$$\begin{aligned} A\,{}^tB &= B\,{}^tA \\ C\,{}^tD &= D\,{}^tC \\ A\,{}^tD - B\,{}^tC &= I. \end{aligned}$$

[10] Let $A = (a_{ij})$ be a $2n \times 2n$ skew-symmetric matrix, the Pfaffian of A, $\mathrm{Pf}(A)$, is defined as: $\mathrm{Pf}(A) = \frac{1}{2^n n!} \sum_{\sigma \in S_{2n}} \mathrm{sgn}(\sigma) \prod_{i=1}^{n} a_{\sigma(21-1),\sigma(2i)}$, where S_{2n} is the symmetric group and $\mathrm{sgn}(\sigma)$ is the signature of the permutation σ.

More abstractly, the symplectic group can be defined as the group of linear transformations of a $2n$−dimensional vector space over a field F which preserve a nondegenerate, skew-symmetric, bilinear form.

When the symplectic matrices take value in $\mathbb{Z}$ we will write $\Gamma_g := \mathrm{Sp}(2g, \mathbb{Z})$. Even though $\mathbb{Z}$ is not a field Γ_g is a group and we define its congruence subgroup of level n as

$$\Gamma_g(n) := \{M \in \Gamma_g : M \equiv I \bmod n\}. \tag{25}$$

These are normal subgroups of Γ_g. A subgroup H of a group G is normal if it is invariant under conjugation, i.e. $ghg^{-1} \in H$ for any $g \in G$ and $h \in H$. If $M \in \Gamma_g$ and $N \in \Gamma_g(n)$ we have that MNM^{-1} is again in $\Gamma_g(n)$ because the matrix elements outside the diagonal are $\frac{k \times 0 \bmod n}{\det M}$ that is again $0 \bmod n$ for all $k \in \mathbb{Z}$ and the matrix elements on the diagonal are $\frac{\det(M) \times 1 \bmod b + k \times 0 \bmod n}{\det(M)}$, for a certain $k \in \mathbb{Z}$, and this is again $1 \bmod n$, thus $\Gamma_g(n)$ is a normal subgroup of Γ_g.

The case $n = 2$ is of particular interest for the applications to string theory:

$$\Gamma_g(2) = \ker(\Gamma_g := \mathrm{Sp}(2g, \mathbb{Z}) \longrightarrow \mathrm{Sp}(2g) := \mathrm{Sp}(2g, \mathbb{Z}_2)), \tag{26}$$

where $\mathbb{Z}_2$ is the field with two elements. The reduction mod two map, as proved by Igusa, is surjective, so that we have $\mathrm{Sp}(2g) \cong \Gamma_g/\Gamma_g(2)$. This group is clearly finite and its order is [J]: $|\mathrm{Sp}(2g)| = 2^{2g-1}(2^{2g}-1)|\mathrm{Sp}(2g-2)| = 2^{2g-1}(2^{2g}-1)2^{2g-3}(2^{2g-2}-1)\cdots 2(2^2-1)$.

3.2. The Action of $\mathrm{Sp}(2g)$ on the Theta Characteristics

In order to construct string amplitudes we will use certain special functions called *theta functions with characteristic*. There is a natural action of the symplectic group on such theta functions. This can be used to construct a special class of functions called modular forms. Before introducing theta functions and modular forms we define here theta characteristics and study the action of the symplectic group on them.

The finite field $\mathbb{Z}_2^{2g}$ has 2^{2g} elements called *period characteristics* and the group Γ_g naturally acts linearly on them through its quotient $\mathrm{Sp}(2g) = \Gamma_g/\Gamma_g(2)$, the action being simply given by the matrix product on the column vectors of $\mathbb{Z}_2^{2g}$. A *theta g-characteristic*, or simply a *theta characteristic*, is a $2 \times g$ array $\begin{bmatrix} a_1 & \cdots & a_g \\ b_1 & \cdots & b_g \end{bmatrix}$, with $b_i, a_i \in \mathbb{Z}_2$. We can define a (non linear) action of the symplectic group on the theta characteristics. Following the abstract definition of Section 3.1., the group $\mathrm{Sp}(2g)$ fixes a nondegenerate skew-symmetric form E on the $\mathbb{Z}_2$-vector space $V = \mathbb{Z}_2^{2g}$. Choosing a symplectic basis for V, which is a basis $e_1, \cdots, e_{2g}$ of V such that $E(e_1, e_j) = 0$ unless $|i-j| = g$ and then $E(e_i, e_j) = 1$, we obtain:

$$\begin{aligned} &E : V \times V \longrightarrow \mathbb{Z}_2, \\ &E(v, w) := v_1 w_{g+1} + \cdots + v_g w_{2g} + v_{g+1} w_1 + \cdots + 2_{2g} w_g, \end{aligned} \tag{27}$$

where $v, w \in \mathbb{Z}_2^{2g}$. More compactly, we can write $E((v', v''), (w', w'')) = {}^t v' w'' + {}^t v'' w'$ and occasionally we will write $v = \binom{v'}{v''}$, where v', v'' are then considered as row vectors.

Let us consider the quadratic form q on the vector space V whose associated bilinear form is E. This is the map:

$$q : V \longrightarrow \mathbb{Z}_2, \qquad q(v+w) = q(v) + q(w) + E(v,w). \tag{28}$$

It is not hard to verify that for each choice of a_i and $b_i \in \mathbb{Z}_2$ the function

$$\begin{aligned} q(v) &= v_1 v_{g+1} + v_2 v_{g+2} + \cdots + v_g v_{2g} + a_1 v_1 + \cdots + a_g v_g \\ &\quad + b_1 v_{g+1} + \cdots + b_g v_{2g} \end{aligned} \tag{29}$$

satisfies $q(v+w) = q(v) + q(w) + E(v,w)$ and that any quadratic form associated to E is of this form. With the compact notations, we can write $q(v) = {}^t v' v'' + av' + bv''$, with row vectors $a = (a_1, \cdots, a_g)$ and $b = (b_1, \cdots, b_g)$. We are now able to give the precise definition of theta characteristics. The *theta characteristic* Δ_q *associated to the quadratic form* q is defined as:

$$\Delta_q := \begin{bmatrix} a_1\, a_2 \ldots a_g \\ b_1\, b_2 \ldots b_g \end{bmatrix} = \begin{bmatrix} a \\ b \end{bmatrix}. \tag{30}$$

To introduce the notion of parity of a theta characteristic we define $e(\Delta_q) := (-1)^{\sum_{i=1}^{g} a_i b_i} \in \{1, -1\}$ and we say that Δ_q is even if $e(\Delta_q) = +1$ and odd elsewhere. One can verify that:

$$e(\Delta_q) 2^g = \sum_{v \in V} (-1)^{q(v)}. \tag{31}$$

It follows that $q(v)$ has $2^{g-1}(2^g+1)$ zeroes in V if Δ_q is even and has $2^{g-1}(2^g-1)$ zeroes if Δ_q is odd. To show this consider an even characteristic Δ_q, from (31) we obtain $2^g = z - p$, where z is the number of v for which $q(v)$ is $0 \pmod 2$ and p is the number of v for which $q(v)$ is $1 \pmod 2$. Hence, we obtain the two equations:

$$\begin{aligned} 2^g &= z - p \\ 2^{2g} &= z + p, \end{aligned}$$

solving this system one obtains the number of zeroes for $q(v)$ for an even Δ_q. The case Δ_q odd is similar but the first equation of the system becomes $-2^g = z - p$.

For any genus g there are 2^{2g} theta characteristics of which $2^{g-1}(2^g+1)$ are even and $2^{g-1}(2^g-1)$ odd. This can be shown as follows ([RF], Chapter 1, Theorem 1). In genus one there are three even characteristics: $\begin{bmatrix} 0 \\ 0 \end{bmatrix}$, $\begin{bmatrix} 1 \\ 0 \end{bmatrix}$, $\begin{bmatrix} 0 \\ 1 \end{bmatrix}$ and one odd $\begin{bmatrix} 1 \\ 1 \end{bmatrix}$. If one borders an even (odd) $(g-1)$-characteristic on the right by an even 1-characteristic, one gets a even (odd) g-characteristic. Instead, bordering an even (odd) $(g-1)$-characteristic by $\begin{bmatrix} 1 \\ 1 \end{bmatrix}$ gives an odd (even) g-characteristic. Thus, if e_i and o_i are the cardinals of the even and odd i-characteristics respectively, then:

$$\begin{aligned} e_g &= 3e_{g-1} + o_{g-1} \\ o_g &= 3o_{g-1} + e_{g-1}. \end{aligned} \tag{32}$$

Adding the two equations, considering that $e_1 + o_1 = 2^2$, and by induction one obtains:

$$e_g + o_g = 2^2(e_{g-1} + o_{g-1}) = 2^{2g}, \tag{33}$$

this prove that in genus g there are exactly 2^{2g} characteristics. Subtracting the two equations (32) and again by induction we obtain:

$$e_g - o_g = 2(e_{g-1} - o_{g-1}) = 2^g. \tag{34}$$

Solving (33) and (34) for e_g and o_g we obtain:

$$\begin{aligned} e_g &= 2^{g-1}(2^g + 1) \\ o_g &= 2^{g-1}(2^g - 1). \end{aligned}$$

The group $\mathrm{Sp}(2g)$ acts naturally on the characteristics by:

$$(g \cdot q)(v) := q(g^{-1}v), \qquad (g \in \mathrm{Sp}(2g),\ v \in V). \tag{35}$$

This action is transitive on both the set of even and odd characteristics which are the two orbits of the action.

3.3. Modular Forms

We recall here that the Siegel upper half space $\mathbb{H}_g$ is the space of complex $g \times g$ symmetric matrices with positive imaginary part. We can see $\mathbb{H}_g$ as a higher dimensional generalization of the half upper complex plane (i.e. the set of complex numbers with positive imaginary part):

$$\mathbb{H}_g := \{\tau \in M_g(\mathbb{C}) : \ {}^t\tau = \tau,\ \mathrm{Im}(\tau) > 0\}. \tag{36}$$

A Siegel modular form of genus g, weight k and level n is a holomorphic function on the Siegel upper half space such that:

$$f : \mathbb{H}_g \longrightarrow \mathbb{C}, \qquad f(M \cdot \tau) = \det(C\tau + D)^k f(\tau) \quad \forall M \in \Gamma_g(n), \tag{37}$$

plus, for $g = 1$, the requirement that f is holomorphic at ∞. The action of Γ_g on the Siegel upper half space is given by

$$M \cdot \tau := (A\tau + B)(C\tau + D)^{-1}, \qquad M := \begin{pmatrix} A & B \\ C & D \end{pmatrix} \in Sp(2g, \mathbb{Z}), \tag{38}$$

for $\tau \in \mathbb{H}_g$. The set of all Siegel modular forms of genus g, weight k and level n form a finite dimensional complex vector space denoted $M_k(\Gamma_g(n))$. For the applications to string theory we are mainly interested to the case $n = 2$. The finite group $Sp(2g)$ has a representation

$$\rho \equiv \rho_k : Sp(2g) \longrightarrow GL(M_k(\Gamma_g(2)))$$

on this vector space defined by

$$(\rho(g^{-1})f)(\tau) := \det(C\tau + D)^{-k} f(M \cdot \tau), \tag{39}$$

where $M \in \Gamma_g$ is any representative of the equivalence class of $g \in Sp(2g)$ and $f \in M_k(\Gamma_g(2))$ (note that $\det(C\tau + D)^{-k} f(M \cdot \tau) = f(\tau)$ for $M \in \Gamma_g(2)$, thus the action of $M \in \Gamma_g$ factors over $\Gamma_g/\Gamma_g(2) = Sp(2g)$). The equality $\rho(gh) = \rho(g)\rho(h)$ for $g, h \in Sp(2g)$ follows from $(MN) \cdot \tau = M \cdot (N \cdot \tau)$ and $\gamma(MN, \tau) = \gamma(M, N \cdot \tau)\gamma(N, \tau)$ where $\gamma(M, \tau) := \det(C\tau + D)$. This shows that ρ effectively defines a group representation.

3.4. Theta Constants with Characteristic

A powerful tool to determine modular forms on $\Gamma_g(2)$ is provided by theta constants with characteristic. Let $\Delta = \left[\begin{smallmatrix} a \\ b \end{smallmatrix}\right]$ be an even characteristic, then one defines a function, called *theta constant*, on the Siegel space $\mathbb{H}_g$ by

$$\theta\left[\begin{smallmatrix} a \\ b \end{smallmatrix}\right](\tau) := \sum_{m \in \mathbb{Z}^g} e^{\pi i((m+a/2)\tau\, {}^t(m+a/2)+(m+a/2)\, {}^tb)} \tag{40}$$

so m is a row vector and $\sum a_i b_i \equiv 0 \bmod 2$. Then, for all $M \in Sp(2g, \mathbb{Z})$, one has ([I1], V.1, Corollary):

$$\theta[M \cdot \Delta](M \cdot \tau) = \kappa(M) e^{2\pi i \phi_\Delta(M)} \det(C\tau + D)^{1/2} \theta[\Delta](\tau), \tag{41}$$

with:

$$\begin{aligned} \phi_\Delta(M) = \sum_{k,l=1}^{g} \tfrac{-1}{8}\Big(({}^tDB)_{kl}a_k a_l - 2({}^tBC)_{kl}a_k b_l + ({}^tCA)_{kl}b_k b_l\Big) + \\ \tfrac{1}{4}(({}^tD)_{kl}a_k - ({}^tC)_{kl}b_k)(A\,{}^tB)_{ll}, \end{aligned}$$

and $\kappa(M)$ a constant independent on the characteristic. Here the action of $M \in \mathrm{Sp}(2g, \mathbb{Z})$ on the characteristic Δ is given by (23). The formula (41) is called *transformation formula*. This formula is explicit except for the constant $\kappa(M)$. See [RF] for the expression of $\kappa(M)$ or, in case of squared theta constants, see [I1]. Note that

$$\theta[\Delta](M^{-1} \cdot \tau) = \theta[M^{-1}M \cdot \Delta](M^{-1} \cdot \tau) = c_{M^{-1},\Delta,\tau}\theta[M \cdot \Delta](\tau) \tag{42}$$

where $c_{M^{-1},\Delta,\tau}$ collects the non-relevant part. Thus the action of M basically maps $\theta[\Delta]$ to $\theta[M \cdot \Delta]$.

The action of Γ_g on the theta characteristics defined in (23) corresponds to its action on the quadratic forms on V defined in Section 3.2.. To show this we have to proof explicitly that

$$(M \cdot q_\Delta)(v) = q_{M \cdot \Delta}(v). \tag{43}$$

From the definition of the action of $\mathrm{Sp}(2g)$ on the quadratic forms $(M \cdot q_\Delta)(v) = q_\Delta(M^{-1}v)$ then we must verify the relation $q_\Delta(M^{-1}v) = q_{M\cdot\Delta}(v)$. From the definition of Γ_g we have $ME\,{}^tM = E$ and taking the inverse of both sides we obtain $M^{-1} = -E\,{}^tME$, so:

$$M^{-1}v = \begin{pmatrix} {}^tD & -{}^tB \\ -{}^tC & {}^tA \end{pmatrix} \begin{pmatrix} v' \\ v'' \end{pmatrix} = \begin{pmatrix} {}^tDv' - {}^tBv'' \\ -{}^tCv' + {}^tAv'' \end{pmatrix}. \tag{44}$$

Using $q_\Delta(v) = q_{\left[\begin{smallmatrix} a \\ b \end{smallmatrix}\right]}(\left(\begin{smallmatrix} v' \\ v'' \end{smallmatrix}\right)) = {}^tv'v'' + av' + bv''$, we get

$$\begin{aligned} q_\Delta(M^{-1}v) = {}&{}^t({}^tDv' - {}^tBv'')(-{}^tCv' + {}^tAv'') \\ &+ a({}^tDv' - {}^tBv'') + b(-{}^tCv' + {}^tAv''). \end{aligned} \tag{45}$$

The non-linear part is

$${}^tv'(-D\,{}^tC)v' + {}^tv'(D\,{}^tA)v'' + {}^tv''(B\,{}^tC)v' - {}^tv''(B\,{}^tA)v''. \tag{46}$$

As M is symplectic, $D\,{}^tC$ is symmetric, hence only the terms $(D\,{}^tC)_{ii}(v'_i)^2$ survives mod 2. But $(v'_i)^2 \equiv v'_i$ mod 2 and thus ${}^tv'(-D\,{}^tC)v' \equiv (D\,{}^tC)_0 v'$ mod 2. Similarly, $v''(B\,{}^tA)v'' \equiv (B\,{}^tA)_0 v''$ mod 2. Next, $B\,{}^tC \equiv I + A\,{}^tD$ mod 2 so that ${}^tv'(D\,{}^tA)v'' + {}^tv''(B\,{}^tC)v' \equiv {}^tv'v''$ mod 2. Thus, we find that

$$\begin{aligned} q_\Delta(M^{-1}v) = \; & {}^tv'v'' + (a\,{}^tD - b\,{}^tC + (D\,{}^tC)_0)v' \\ & + (-a\,{}^tB + b\,{}^tA + (B\,{}^tA)_0)v'' \end{aligned} \tag{47}$$

so $q_\Delta(M^{-1}v) = q_{M\cdot\Delta}(v)$, as desired. This clarifies the definition of the action given in (23).

3.5. The Subgroup $O^+(2g)$ of $\mathrm{Sp}(2g)$

The stabilizer subgroup of the (even) characteristic $[0] := \left[\begin{smallmatrix}0\ldots0\\0\ldots0\end{smallmatrix}\right]$ is the subgroup $\Gamma_g(1,2)$, a special case of a series of subgroups of Γ_g:

$$\Gamma_g(n,2n) := \{M \in \Gamma_g(n) : \mathrm{diag} A\,{}^tB \equiv \mathrm{diag} C\,{}^tD \equiv 0 \bmod 2n\}. \tag{48}$$

In case n is even $\Gamma_g(n,2n)$ is a normal subgroup of $\Gamma_g \equiv \Gamma_g(1)$. We call $O^+(2g)$ the image of $\Gamma_g(1,2)$ in $\mathrm{Sp}(2g)$

$$O^+(2g) := \Gamma_g(1,2)/\Gamma_g(2) \qquad (\subset Sp(2g)). \tag{49}$$

As $Sp(2g)$ acts transitively on the even theta characteristics, there is a natural bijection

$$Sp(2g)/O^+(2g) \longrightarrow \{\Delta : \Delta \text{ even}\}, \qquad hO^+(2g) \longmapsto h\cdot[0], \tag{50}$$

for $h \in \mathrm{Sp}(2g)$. In particular, $[Sp(2g) : O^+(2g)] = 2^{g-1}(2^g+1)$. One has $O^+(2) \cong \mathbb{Z}/2\mathbb{Z}$ and $O^+(4)$ is isomorphic to the subgroup of the symmetric group $S_6 \cong Sp(4)$ consisting of all permutations σ such that $\sigma(\{1,2,3\}) \subset \{1,2,3\}$ or $\sigma(\{1,2,3\}) = \{4,5,6\}$. Thus $S_3 \times S_3$ is a subgroup of index two in $O^+(4)$ and $|O^+(4)| = (3!)\cdot(3!)\cdot 2 = 72$. One has $O^+(6) \cong S_8$, the symmetric group of order $8!$ and $O^+(8)$ is the quotient of the subgroup of elements of[11] $W(E_8)$ with determinant $+1$ in the standard 8-dimensional representation, by its center, generated by $-I$.

3.6. Theta Constants and the Heisenberg Group

In this subsection we will study the relation between modular forms and theta constants. We will see that the modular group acts on the theta constant projectively instead of linearly. Recognizing the action of a finite Heisenberg group will help us to obtain modular forms from suitable polynomials in theta constants.

The theta constants are *almost modular* forms of weight $1/2$ on $\Gamma_g(4,8)$; due to the presence of the constant $\kappa(M)$ we used the expression "almost modular". This can be shown by using the "transformation formula"

$$\begin{aligned} \theta[\Delta](M\cdot\tau) &= \theta[MM^{-1}\Delta](M\cdot\tau) \\ &= \kappa(M)e^{2\pi i\Phi_\Delta(M)}\det(C\tau+D)^{1/2}\theta[M^{-1}\Delta](\tau). \end{aligned}$$

[11] The Weyl group of E_8.

The exponential phase takes the value 1 if $\Phi_\Delta(M)$ is an integer. As $M \in Sp(2g,\mathbb{Z})$, also $M^{-1} \in Sp(2g,\mathbb{Z})$, thus M satisfies also ${}^tBD - {}^tDB = 0$ and ${}^tAC - {}^tCA = 0$, that mean that ${}^tDB$ and ${}^tCA$ are symmetric matrices. Hence, the integers $a_k a_l, b_k b_l$ in $a^tDB^ta + bCA^tb$ are multiplied by an even integer if $k \neq l$, so that they do not contribute to the exponential if ${}^tDB$ and ${}^tCA$ are 0 mod 4. For $k = l$ we have $\sum_k(a_k^2({}^tDB)_{kk} + b_k^2({}^tCA)_{kk})$, but note that $a_k^2 \equiv a_k \bmod 2$. For a $g\times g$ matrix M, let $\mathrm{diag}(M)$ be the column vector $(M_{11}, M_{22}, \ldots, M_{gg})$ of diagonal entries. Then, the last term is $a\,\mathrm{diag}({}^tDB) + b\,\mathrm{diag}({}^tCA)$ which does not contribute to the phase if its value is $0 \bmod 8$. These two requests are precisely the conditions defining the subgroup $\Gamma_g(4,8)$. Note that if $M \in \Gamma_g(4,8)$ the term $2({}^tBC)_{kl}a_k b_l$ is a multiple of eight so that it does not contribute to the phase and the same hold true for the second term in $\Phi_\Delta(M)$ because A^tB is $0 \bmod 8$. Moreover $M^{-1}\Delta = \Delta$ for all $M \in \Gamma_g(2)$ (or in some its subgroup). Just the constant $\kappa(M)$ survives.

To determine the modular forms of even weight on $\Gamma_g(2)$ it is convenient to define the 2^g (second order) theta constants:

$$\Theta[\sigma](\tau) := \theta[\begin{smallmatrix}\sigma\\0\end{smallmatrix}](2\tau), \qquad [\sigma] = [\sigma_1\,\sigma_2\,\ldots\,\sigma_g],\ \sigma_i \in \{0,1\},\ \tau \in \mathbb{H}_g.$$

These theta constants, being evaluated in 2τ, are almost modular forms of weight 1/2 on $\Gamma_g(2,4)$:

$$\Theta[\sigma](M\cdot\tau) = \theta[\begin{smallmatrix}\sigma\\0\end{smallmatrix}]\left(2\frac{A\tau+B}{C\tau+D}\right) = \theta[\begin{smallmatrix}\sigma\\0\end{smallmatrix}](\tilde{M}\cdot 2\tau),$$

with $\tilde{M} = \left(\begin{smallmatrix}A & 2B\\ \frac{C}{2} & D\end{smallmatrix}\right)$ and similar considerations as the previous ones lead to the right conclusion. As we shall see, the invariants of degree $4k$ of the quotient group $\Gamma_g(2)/\Gamma_g(2,4) \cong \mathbb{Z}_2^{2g}$ in the ring of polynomials in the $\Theta[\sigma]$'s are modular forms of weight $2k$ on $\Gamma_g(2)$. However, the quotient group $\Gamma_g(2)/\Gamma_g(4,2)$ doesn't act linearly on the $\Theta[\sigma](\tau)$. Using the action ρ defined in (39) we obtain that $\rho(g_N^{-1})\rho(g_M^{-1}) \neq \rho(g_{(MN)^{-1}})$ for the presence in the exponential of the phases $\Phi_\Delta(M) + \Phi_\Delta(N)$ which are not equal to $\Phi_\Delta(NM)$. Here g_M, g_N stand for the equivalence classes in $\Gamma_g(2)/\Gamma_g(2,4)$ of the matrices $M,\ N \in \Gamma_g(2)$ respectively. Thus, a term depending on the characteristic Δ remains. Moreover, $\kappa(M)\kappa(N) \neq \kappa(MN)$, but we will see how to get rid of this term. The previous considerations lead to take into account a central extension of the quotient group $\Gamma_g(2)/\Gamma_g(2,4)$, the Heisenberg group.

The finite Heisenberg group is defined as $H_g = \mu_4 \times \mathbb{Z}_2^g \times \mathbb{Z}_2^g$, where $\mu_4 = \{z \in \mathbb{C} : z^4 = 1\}$ is the multiplicative group of fourth roots of unity. The group composition law is $(s,x,u)(t,y,v) = (st(-1)^{uy}, x+y, u+v)$, with $uy = u_1y_1 + \cdots + u_gy_g$ for $u = (u_1,\cdots u_g)$, $y = (y_1,\cdots y_g) \in \mathbb{Z}_2^g$. The center of H_g is the multiplicative group μ_4 and the quotients H_g/μ_4 are isomorphic to $\mathbb{Z}_2^{2g}$. Let us consider the ring of polynomials in 2^g variables X_σ, where $[\sigma] = [\sigma_1\,\sigma_2\,\cdots\,\sigma_g]$, $\sigma_i \in \{0,1\}$. On this space we can define the (Schrödinger) representation of the Hesienberg group

$$(s,x,u)X_\sigma := s(-1)^{(x+\sigma)u}X_{\sigma+x},$$

and this action is extended to polynomials in X_σ's in an obvious way. Directly from this definition, it follows that a polynomial in the X_σ is invariant under the center μ_4 of the Heisenberg group if and only if its degree is a multiple of four, so the H_g invariant polynomials result to be modular forms of even weight. We will denote the subring of these invariants in the ring of polynomials in X_σ as $\mathbb{C}[\cdots, X_\sigma, \cdots]^{H_g}$. The action of H_g on the ring of polynomials $\mathbb{C}[\cdots, X_\sigma, \cdots]$ induces an action on the ring of the (second order) theta constants $\mathbb{C}[\cdots, \Theta[\sigma], \cdots]$ under the map $X_\sigma \mapsto \Theta[\sigma]$. The subspace of $M_{2k}(\Gamma_g(2))$ of the Heisenberg invariants is denoted by $M^\theta_{2k}(\Gamma_g(2)) \subset M_{2k}(\Gamma_g(2))$, where $M^\theta_{2k}(\Gamma_g(2)) := \mathbb{C}[\cdots, \Theta, \cdots]^{H_g}_{2k}$.

Using the two generators $(1, x, 0)$ and $(1, 0, u)$ of the Heisenberg group it is not hard to construct a basis for the space of the invariants. We fixed 1 as element of the center because the latter acts trivially on polynomials of degree four. First, each monomial in an Heisenberg invariant polynomial P is a product $\prod_{i=1}^{4n} X_{\sigma_i}$ which must be invariant under the action of $(1, 0, u)$. This means $(1, 0, u)\prod_{i=1}^{4n} X_{\sigma_i} = (-)^{(\sum_i \sigma_i)u}\prod_{i=1}^{4n} X_{\sigma_i}$, which is invariant for all u if $\sum \sigma_i = 0$. Next, P must be invariant also for the elements of the form $(1, x, 0)$, which means that all monomials in P of the type $\prod_{i=1}^{4n} X_{\sigma_i + x}$, for any $x \in \mathbb{Z}_2^g$, must have the same coefficient in P. Thus a basis for the subring of the Heisenberg invariants is provided by polynomials of the form $\sum_x \prod_{i=1}^{4n} X_{\sigma_i + x}$, where $\sum \sigma = 0$.

3.6.1. Transvections

The group $Sp(2g)$ is generated by transvections t_v, for $v \in V$, which are analogous to reflections in orthogonal groups ([J], § 6.9). They are defined as:

$$t_v : V \longrightarrow V, \qquad t_v(w) := w + E(w, v)v.$$

It is straightforward to verify that $t_v \in Sp(2g)$. In fact the same formula works also for $\mathbb{Z}$ in place of $\mathbb{Z}_2$ and then defines elements in $Sp(2g, \mathbb{Z})$. As $gt_vg^{-1} = t_{g(v)}$ for $g \in Sp(2g, \mathbb{Z}_2)$ and $v \in V$, the non-trivial transvections form a conjugacy class. It is not hard to prove that t_v is an involution, i.e. $t_v^2 = 1$.

Let us now determine how tranvections act on the characteristics. Let $v \in V$ and let q be a quadratic form with associated bilinear form E and characteristic Δ_q. As t_v is an involution, $q(v + w) = q(v) + q(w) + E(v, w)$ for all $v, w \in V$ and $q(av) = aq(v)$ for $a \in \mathbb{Z}_2$, we have

$$(t_v \cdot q)(w) = q(t_v(w)) - q(w + E(v, w)v) = q(w) + E(v, w)q(v) + E(v, w)^2.$$

Hence we get the simple rule:

$$(t_v \cdot q)(-) = \begin{cases} q(-) & \text{if } \ q(v) = 1, \\ q(-) + E(v, -) & \text{if } \ q(v) = 0, \end{cases}$$

so $t_{\binom{v'}{v''}} \cdot q_{\left[\substack{a \\ b}\right]} = q_{\left[\substack{a+v'' \\ b+v'}\right]}$ in case $q_{\left[\substack{a \\ b}\right]}(\binom{v'}{v''}) = 0$.

Using the transvections we obtain the action of $Sp(2g)$ on the Heisenberg invariants in a simple manner and a computer can be used to perform all computations (see the Appendix B of [DvG] for an exhaustive discussion about transvections).

3.6.2. Dimension of the space of Heisenberg invariants

We will show that the functions we need to construct superstrings amplitudes belong to the subspace of the Heisenberg invariants given by the vector space $\mathbb{C}[\cdots, X_\sigma, \cdots]_n$ of homogeneous polynomials of degree $n = 16$ in the X_σ's. Here we show that the dimensions of these spaces are given by the formula:

$$\dim(\mathbb{C}[\ldots, X_\sigma, \ldots]_{4n})^{H_g} = 2^{-2g}\left(\binom{2^g + 4n - 1}{4n} + (2^{2g} - 1)\binom{2^{g-1} + 2n - 1}{2n}\right).$$

We list in Table 2 such dimensions for the lowest values of genus g and degree $4n$.

Table 2. Dimension of some space of Heisenberg invariants.

g / degree	4	8	12	16
1	2	3	4	5
2	5	15	35	69
3	15	135	870	3993
4	51	2244	69615	1180396

To prove the formula we will employ the theory of finite group representations. For fixed g let ρ_n be the representation of the Heienberg group on the vector space of homogeneous polynomials in X_σ's, as introduced before:

$$\rho_n : H_g \to GL(\mathbb{C}[\cdots, X_\sigma, \cdots]_n]). \tag{51}$$

Clearly, the Heisenberg invariants are the space of the trivial subrepresentation of ρ_n. Thus its dimension is

$$\dim \mathbb{C}[\cdots, X_\sigma \cdots]_n^{H_g} = \langle \rho_n, 1_{H_g} \rangle_{H_g}, \tag{52}$$

i.e. the multiplicity of the trivial representation 1_{H_g} of H_g in ρ_n. The scalar product of the characters is given by $\langle \rho_n, 1_{H_g} \rangle_{H_g} = \frac{1}{|H_g|} \sum_{h \in H_g} \mathrm{Tr}(\rho_n(h))$, where tr is the trace and $|H_g|$ is the number of elements of the group H_g. Consider the element $(1, x, u) \in \mathbb{H}_g$. If $xu = 0$, but $(x, u) \neq (0, 0)$ then $(1, x, u)$ has order two in $\mathbb{H}_g$ and the eigenvalues of (t, x, u) on $\mathbb{C}[\ldots, X_\sigma, \ldots]_1$ are t and $-t$, each with multiplicity 2^{g-1} for all $t \in \mu_4$.

If $\alpha_1, \ldots, \alpha_N$, $N = 2^g$ are the eigenvalues of (t, x, u) on $\mathbb{C}[\ldots, X_\sigma, \ldots]_1$, the eigenvalues on $\mathbb{C}[\ldots, X_\sigma, \ldots]_n$ are the $\alpha_1^{m_1} \alpha_2^{m_2} \ldots \alpha_N^{m_N}$ with $\sum m_i = n$. As the trace is the sum of the eigenvalues, we get with a variable X:

$$\sum_n \mathrm{Tr}(\rho_n(t, x, u)) X^n = \prod_{i=1}^{N} (1 - \alpha_i X)^{-1}.$$

So if $(x, u) \neq (0,0)$ we have

$$\sum_n \mathrm{Tr}(\rho_n(t,x,u))X^n = (1 - i^{2a}X^2)^{-2^{g-1}}$$
$$= \sum_m (-1)^{am}\binom{2^{g-1}+m-1}{m} X^{2m},$$

and in case $(x,u) = (0,0)$ the trace is just

$$\sum_n \mathrm{Tr}(\rho_n(t,0,0))X^n = \sum_n t^n(\dim \mathbb{C}[\ldots, X_\sigma, \ldots]_n)X^n$$
$$= \sum_n t^n \binom{2^g+n-1}{n} X^n.$$

Thus, we obtain the anticipated formula:

$$\dim(\mathbb{C}[\ldots, X_\sigma, \ldots]_{4n})^{H_g} = 2^{-2g}\left(\binom{2^g+4n-1}{4n} + (2^{2g}-1)\binom{2^{g-1}+2n-1}{2n}\right),$$

note that non-trivial invariants have degree multiple of 4.

3.6.3. The ring of modular forms

Let us call $M^\theta_{2k}(\Gamma_g(2))$ the spaces of the Heisenberg invariants of degree $4k$ (and weight $2k$) in the second order theta functions. These are the images under the surjective maps

$$\begin{aligned} \mathbb{C}[\ldots, X_\sigma, \ldots]^{H_g}_{4k} &\longrightarrow M^\theta_{2k}(\Gamma_g(2)) := \mathbb{C}[\ldots, \Theta[\sigma], \ldots]^{H_g}_{4k}, \\ X_\sigma &\longmapsto \Theta[\sigma] \end{aligned}$$

of the Heisenberg invariant polynomials of degree $4k$. These maps define a surjective $\mathbb{C}$-algebra homomorphism

$$\mathbb{C}[\ldots, X_\sigma, \ldots]^{H_g} \longrightarrow M^\theta(\Gamma_g(2)) := \oplus_k M^\theta_{2k}(\Gamma_g(2)), \tag{53}$$

whose kernel is the ideal of algebraic relations between the $\Theta[\sigma]$'s. This means that a polynomial $F(\cdots, X_\sigma, \cdots)$ maps to zero if and only if

$$F(\cdots, \Theta[\sigma](\tau), \cdots) = 0$$

for all $\tau \in \mathbb{H}_g$. For $g = 1, 2$ there are no polynomials vanishing on the image. In case $g = 3$ there is a homogeneous polynomial F_{16}, of degree sixteen in eight variables, vanishing on the image [vGvdG], so that $M^\theta(\Gamma_g(2)) = \mathbb{C}[\cdots, X_\sigma, \cdots]^{\mathbb{H}_g}/(F_{16})$. For $g \geq 4$ there are many algebraic relations between the $\Theta[\sigma]$'s, but a complete description of these relations is not known. The graded ring of modular forms of even weight on $\Gamma_g(2)$ is the normalization of the ring[12] of the $\Theta[\sigma]$'s (cf. [SM1] Thm 2, [R1], [R2]):

$$\oplus_{k=0}^{\infty} M_{2k}(\Gamma_g(2)) = (\mathbb{C}[\ldots,\Theta[\sigma],\ldots]^{H_g})^{Nor}.$$

In case $g = 1, 2$ there are no relations and the rings of invariants are already normal. In case $g = 3$, there is one relation given by $F_{16}(\ldots,\Theta[\sigma],\ldots) = 0$, and it has been shown by Runge ([R1], [R2]) that the quotient of the ring of invariants by the ideal generated by this relation is again normal. This implies that any modular form of weight $2k$ can be written as a homogeneous polynomial of degree $4k$ in the $\Theta[\sigma]$'s if $g \leq 3$, and

$$M^{\theta}_{2k}(\Gamma_g(2)) = M_{2k}(\Gamma_g(2)) \qquad \text{for} \quad g = 1, 2, 3.$$

This polynomial is unique for $g \leq 2$. For $g = 3$ it is unique if its degree is less than 15, otherwise it is unique up to the addition of $F_{16}G_{4k-16}$, where G_{4k-16} is any homogeneous polynomial of degree $4k - 16$ in the $\Theta[\sigma]$'s.

For $g > 3$ there will always be non-trivial relations and if $g > 4$ the ring $\mathbb{C}[\ldots,\Theta[\sigma],\ldots]^{H_g}$'s is not normal, (cf. [OSM], Theorem 6, but note that our H_g is slightly different from their one). In case the ring is not normal, there are also quotients G_{4k+d}/H_d of homogeneous polynomials in the $\Theta[\sigma]$'s, of degree $4k + d$ and d respectively, which are modular forms of weight $4k$ (but which cannot be written as a polynomial in the $\Theta[\sigma]$'s). These observations will play a crucial role in proving the uniqueness of superstring amplitudes. Actually, in genus two and three the proof of uniqueness is based on the result that every modular forms of weight 8 are polynomial in the theta constants, see section 5.2.2. and 5.5.. For the case $g = 4$ we will be able to prove the uniqueness in a weakened form, that is by assuming the polynomiality for the superstring measures 5.7.1..

3.7. Turning back to the classical theta constants with characteristic

To describe the spaces of modular forms $M^{\theta}_{2k}(\Gamma_g(2))$ it is convenient to use also the classical theta functions with arbitrary characteristics $\theta[\Delta]$. Recall that by this we mean that the argument is τ and not 2τ as in the relation defining the $\Theta[\sigma]$. In particular, we are interested in decomposing these spaces into irreducible representations for the group $Sp(2g)$ and we want to describe their subspaces of O^+-invariants as well as O^+-anti-invariants. These will play a crucial role in the construction of superstring measures.

3.7.1. The quadratic relations between the $\theta[\Delta]$'s and the $\Theta[\sigma]$'s

A classical formula for theta functions shows that any product of two $\Theta[\sigma]$'s is a linear combination of the $\theta[\Delta]^2$. Note that there are 2^g functions $\Theta[\sigma]$ and thus there are $(2^g + 1)2^g/2 = 2^{g-1}(2^g+1)$ products $\Theta[\sigma]\Theta[\sigma']$. This is also the number of even characteristics, and the products $\Theta[\sigma]\Theta[\sigma']$ span the same space (of modular forms of weight 1) as the $\theta[\Delta]^2$'s, which has dimension $2^{g-1}(2^g + 1)$ (see [vG2] Lemma (2.7), [RF]).

[12] An Abelian ring A is normal if it does not contains nontrivial nilpotents and is integrally closed w.r.t. its quotient ring $Q[A]$. In other words, any polynomial equation whose coefficients are fractions in $Q[A]$ has solutions in A. A non normal ring can be normalized by takeing its closure in $Q[A]$, see [L].

As the degree of an H_g-invariant homogeneous polynomial in the $\Theta[\sigma]$ is a multiple of four, say $4k$, it can be written as a homogeneous polynomial of degree $2k$ in the $\theta[\Delta]^2$'s. Thus for $g \leq 3$, any element in $M_{2k}(\Gamma_g(2))$ is a homogeneous polynomial of degree $2k$ in the $\theta[\Delta]^2$'s.

The $\theta[\Delta]^2$ are the better known functions and their transformation under $\Gamma_g(1)$ is easy to understand, but the $\Theta[\sigma]$ have the advantage that they are algebraically independent for $g \leq 2$ and there is a unique relation of degree 16 for $g = 3$. In contrast, there are many quadratic relations between the $\theta[\Delta]^2$'s, for example Jacobi's relation in $g = 1$. The classical formula used here is (cf. [I1] IV.1):

$$\theta[{}^a_b]^2 = \sum_\sigma (-1)^{\sigma b} \Theta[\sigma]\Theta[\sigma + a] \tag{54}$$

where we sum over the 2^g vectors $\sigma \in \mathbb{Z}_2^g$ and $[{}^a_b]$ is an even characteristic, so $a\,{}^t b = 0$ $(\in \mathbb{Z}_2)$. These formulae are easily inverted to give:

$$\Theta[\sigma]\Theta[\sigma + a] = \tfrac{1}{2^g} \sum_b (-1)^{\sigma b} \theta[{}^a_b]^2.$$

It is easy to see that the $\theta[\Delta]^2$ span one-dimensional subrepresentations of H_g. Indeed, using the classical formula we find

$$(s, x, u)\theta[{}^a_b]^2 = s^2(-1)^{ua+xb}\theta[{}^a_b]^2.$$

This implies that the $\theta[{}^a_b]^4$ are Heisenberg invariants and thus are in $M_2(\Gamma_g(2))$. More generally, we have:

$$\prod_i^{2k} \theta[{}^{a_i}_{b_i}]^2 \in M_{2k}(\Gamma_g(2)) \quad \text{iff} \quad \sum a_i = \sum b_i = 0\ (\in \mathbb{Z}_2).$$

For example in case $g = 1$ one has

$$\theta[{}^0_0]^2 = \Theta[0]^2 + \Theta[1]^2, \qquad \theta[{}^0_1]^2 = \Theta[0]^2 - \Theta[1]^2, \qquad \theta[{}^1_0]^2 = 2\Theta[0]\Theta[1],$$

or, equivalently,

$$\Theta[0]^2 = (\theta[{}^0_0]^2 + \theta[{}^0_1]^2)/2, \quad \Theta[1]^2 = (\theta[{}^0_0]^2 - \theta[{}^0_1]^2)/2, \quad \Theta[0]\Theta[1] = \theta[{}^1_0]^2/2.$$

Note that upon substituting the first three relations in Jacobi's relation $\theta[{}^0_0]^4 = \theta[{}^0_1]^4 + \theta[{}^1_0]^4$ we obtain a trivial identity in the $\Theta[\sigma]$'s.

3.8. Modular group and representations

Using the classical formula we find that $\theta[0]^4 = (\sum_\sigma \Theta[\sigma]^2)^2$, and it is clear that this functions is Heisenberg invariant and thus defines a modular form of weight 2 on $\Gamma_g(2)$. For $g \in O^+(2g)$ we have $g \cdot [0] = [0]$ and the explicit transformation formula for theta constants shows that $\theta[0]^4$ transforms by a non-trivial character which we denote by ϵ:

$$\rho(g)\theta[0]^4 = \epsilon(g)\theta[0]^4, \qquad \epsilon : O^+(2g) \longrightarrow \{\pm 1\}. \tag{55}$$

For $g \geq 3$, this homomorphism is the only non-trivial one dimensional representation of $O^+(2g)$ and its kernel is a simple group.

3.8.1. Thomae formula and the case $g = 2$

The case $g = 2$ is quite simple and it is interesting to expand some details. For $g = 2$ there are 16 characteristics, six odd and ten even. The odd characteristics are:

$$\nu_1 = \begin{bmatrix} 0 & 1 \\ 0 & 1 \end{bmatrix} \qquad \nu_2 = \begin{bmatrix} 1 & 0 \\ 1 & 0 \end{bmatrix} \qquad \nu_3 = \begin{bmatrix} 0 & 1 \\ 1 & 1 \end{bmatrix}$$

$$\nu_4 = \begin{bmatrix} 1 & 0 \\ 1 & 1 \end{bmatrix} \qquad \nu_5 = \begin{bmatrix} 1 & 1 \\ 0 & 1 \end{bmatrix} \qquad \nu_6 = \begin{bmatrix} 1 & 1 \\ 1 & 0 \end{bmatrix}.$$

The even ones are:

$$\delta_1 = \begin{bmatrix} 0 & 0 \\ 0 & 0 \end{bmatrix} \qquad \delta_2 = \begin{bmatrix} 0 & 0 \\ 0 & 1 \end{bmatrix} \qquad \delta_3 = \begin{bmatrix} 0 & 0 \\ 1 & 0 \end{bmatrix}$$

$$\delta_4 = \begin{bmatrix} 0 & 0 \\ 1 & 1 \end{bmatrix} \qquad \delta_5 = \begin{bmatrix} 0 & 1 \\ 0 & 0 \end{bmatrix} \qquad \delta_6 = \begin{bmatrix} 0 & 1 \\ 1 & 0 \end{bmatrix}$$

$$\delta_7 = \begin{bmatrix} 1 & 0 \\ 0 & 0 \end{bmatrix} \qquad \delta_8 = \begin{bmatrix} 1 & 0 \\ 0 & 1 \end{bmatrix} \qquad \delta_9 = \begin{bmatrix} 1 & 1 \\ 0 & 0 \end{bmatrix}$$

$$\delta_{10} = \begin{bmatrix} 1 & 1 \\ 1 & 1 \end{bmatrix}$$

Note that each even characteristic can be written in two different ways as sum $\pmod 2$ of three odd characteristics and in the sums none odd characteristic is repeated [RF]. For example $\delta_1 = \nu_1 + \nu_4 + \nu_6 = \nu_2 + \nu_3 + \nu_5$. Each set of three odd characteristics that summed gives an even characteristic is called a *triad*. We report all the triads in Table 3. The Thomae formula [Mu, F] allows to express the fourth power of the theta constants in

Table 3. The two triads of odd characteristics giving the same even characteristic.

Triads	$[\delta]$
146 235	$\left[\begin{smallmatrix} 0 & 0 \\ 0 & 0 \end{smallmatrix}\right]$
126 345	$\left[\begin{smallmatrix} 0 & 0 \\ 0 & 1 \end{smallmatrix}\right]$
125 346	$\left[\begin{smallmatrix} 0 & 0 \\ 1 & 0 \end{smallmatrix}\right]$
145 236	$\left[\begin{smallmatrix} 0 & 0 \\ 1 & 1 \end{smallmatrix}\right]$
124 356	$\left[\begin{smallmatrix} 0 & 1 \\ 0 & 0 \end{smallmatrix}\right]$
156 234	$\left[\begin{smallmatrix} 0 & 1 \\ 1 & 0 \end{smallmatrix}\right]$
123 456	$\left[\begin{smallmatrix} 1 & 0 \\ 0 & 0 \end{smallmatrix}\right]$
134 256	$\left[\begin{smallmatrix} 1 & 0 \\ 0 & 1 \end{smallmatrix}\right]$
136 245	$\left[\begin{smallmatrix} 1 & 1 \\ 0 & 0 \end{smallmatrix}\right]$
135 246	$\left[\begin{smallmatrix} 1 & 1 \\ 1 & 1 \end{smallmatrix}\right]$

term of the six branch points of the genus two Riemann surface on which they are defined

$$\theta^4[\delta] = c\,\epsilon_{S,T} \prod_{i,j\in S\; i<j} (u_i - u_j) \prod_{k,l\in T\; k<l} (u_k - u_l) =: \epsilon_{S,T} P_{S,T}, \tag{56}$$

where the u_i's are the six branch points, S and T contain the indices of the two triad of the odd characteristics giving the even characteristic of the theta constant. For example, for δ_4, from Table 3, we have $S = \{1,4,5\}$ and $T = \{2,3,6\}$, $\epsilon_{S,T}$ is a sign depending on the triads and c is a constant independent from the characteristic. Using the many relations between the $\theta[\delta]^4$'s one can determine the relative sign appearing in the Thomae formula: if $\sum_\delta a_\delta \theta[\delta]^4 = 0$ it also holds $\sum_\delta a_\delta \epsilon_{S,T} P_{S,T} = 0$, with the same a_δ. For instance, considering the relation $\theta[\delta_1]^4 - \theta[\delta_4]^4 - \theta[\delta_6]^4 - \theta[\delta_7]^4 = 0$, which can be obtained using the classical formula, we obtain the relative signs between this four $\theta[\delta]^4$'s. Using some other relations we fixe all the relative signs. We report these signs in Table 4. For example,

Table 4. Relative signs between the $\theta[\delta]^4$'s for the thomae formula.

146	126	125	145	124	156	123	134	136	135
235	345	346	236	356	234	456	256	245	246
δ_1	δ_2	δ_3	δ_4	δ_5	δ_6	δ_7	δ_8	δ_9	δ_{10}
-1	1	1	-1	1	-1	1	-1	-1	-1

we have:

$$\theta[\delta_4]^4 = -c(u_1 - u_4)(u_1 - u_5)(u_4 - u_5)(u_2 - u_3)(u_2 - u_6)(u_3 - u_6), \tag{57}$$

and

$$\begin{aligned} \theta[0]^4 &= \theta[\delta_1]^4 \\ &= -c(u_1 - u_4)(u_1 - u_6)(u_4 - u_6)(u_2 - u_3)(u_2 - u_5)(u_3 - u_5). \end{aligned} \tag{58}$$

From the expression of $\theta[0]^4$ obtained with the Thomae formula we conclude that ϵ is the product of the sign character on the subgroup $S_3 \times S_3$ of $O^+(4)$ and $\epsilon(g) = 1$ if $g = (15)(24)(36)$, where we identify $O^+(4)$ with a subgroup of S_6 as in Section 3.5..

3.8.2. The space of O^+-(anti)-invariants

For applications to superstring measures, we will be particularly interested in the subspace of $O^+(2g)$-anti-invariants of weight 6

$$M_6(\Gamma_g(2))^\epsilon := \{f \in M_6(\Gamma_g(2)) \,:\, \rho(g)f = \epsilon(g)f \quad \forall g \in O^+(2g)\}$$

and the space of O^+-invariants of weight 8

$$M_8(\Gamma_g(2))^{O^+} := \{f \in M_8(\Gamma_g(2)) \,:\, \rho(g)f = f \quad \forall g \in O^+(2g)\}.$$

It should be noted that $Sp(2g)$ permutes the $\theta[\Delta]^{4k} \in M_{2k}(\Gamma_g(2))$, up to sign if k is odd. This fact is particular evident in genus two using the Thomae formula and the identification $\mathrm{Sp}(4) \cong S_6$. Thus it is not hard to write down some invariants or anti-invariants, but the problem is to find all of them.

3.8.3. The dimensions of the O^+-(anti)-invariants

To find all O^+-(anti)-invariants we have to know the dimensions of this space. In this section we apply the representation theory of finite groups to $\mathrm{Sp}(2g)$ in order to decompose its representations into irreducibles ones. Once the decomposition of an $\mathrm{Sp}(2g)$-representation into irreducibles is known, it is easy to find the dimension of the O^+-(anti)-invariants. The dimension of the O^+-invariants in V is the multiplicity of the trivial representation $\mathbf{1}$ of O^+ in the O^+-representation $\mathrm{Res}^{Sp}_{O^+}(V)$ (the restriction of the representation from $Sp(2g)$ to $O^+(2g)$):

$$\dim V^{O^+} = \langle\, \mathrm{Res}^{Sp}_{O^+}(V), 1\,\rangle_{O^+} = \langle\, V, Ind^{Sp}_{O^+}(1)\,\rangle_{Sp}$$

where the second equality is Frobenius reciprocity, see Section A. According to Frame [F2] one has:

$$\mathrm{Ind}^{Sp}_{O^+}(1) = \mathbf{1} + \sigma_\theta, \qquad \dim\sigma_\theta = 2^{g-1}(2^g+1) - 1 = (2^g-1)(2^g+2)/2,$$

where $\mathbf{1}$ is the trivial representation and σ_θ is an irreducible representation of $Sp(2g)$. Note that $\dim \mathrm{Ind}^{Sp}_{O^+}(\mathbf{1}) = [Sp(2g) : O^+(2g)] = 2^{g-1}(2^g+1)$. Thus if the multiplicity of $\mathbf{1}$ and σ_θ in V is n_1 and n_θ respectively, then $\dim V^{O^+} = n_1 + n_\theta$.

Similarly, the dimension of the O^+-anti-invariants in V is the multiplicity of the representation ϵ of O^+ in the O^+-representation $\mathrm{Res}^{Sp}_{O^+}(V)$:

$$\dim V^\epsilon = \langle \mathrm{Res}^{Sp}_{O^+}(V), \epsilon\,\rangle_{O^+} = \langle V, \mathrm{Ind}^{Sp}_{O^+}(\epsilon)\,\rangle_{Sp}.$$

According to Frame [F2], the induced representation has two irreducible components:

$$\mathrm{Ind}^{Sp}_{O^+}(\epsilon) = \rho_\theta \oplus \rho_r, \qquad \begin{cases} \dim\rho_\theta &= (2^g+1)(2^{g-1}+1)/3, \\ \dim\rho_r &= (2^g+1)(2^g-1)/3. \end{cases}$$

Thus if the multiplicity of ρ_θ and ρ_r in V are n_θ and n_r respectively, then $\dim V^\epsilon = n_\theta + n_r$.

3.8.4. The $\mathrm{Sp}(2g)$-representation on $M_2^\theta(\Gamma_g(2))$ and on $\mathrm{Sym}^2(M_2^\theta(\Gamma_g(2)))$

The representation ρ_2 of $Sp(2g)$ on the subspace $M_2^\theta(\Gamma_g(2)) \subset M_2(\Gamma_g(2))$ was shown to be irreducible and isomorphic to ρ_θ by van Geemen in [vG2].

Frame has shown that the $Sp(2g)$-representation $\mathrm{Sym}^2(M_2^\theta(\Gamma_g(2)))$ decomposes into irreducible representations as follows:

$$\mathrm{Sym}^2(\rho_\theta) = \mathbf{1} + \sigma_\theta + \sigma_c, \qquad \dim\sigma_c = 2^{g-2}(2^g+1)(2^g-1)(2^g+2), \tag{59}$$

and σ_θ as in 3.8.3.. The functions $\theta[\Delta]^8$ are in $M_4^\theta(\Gamma_g(2))$. They are permuted (without signs) by $Sp(2g)$ and span the subrepresentation $\mathbf{1} + \sigma_\theta$. The trivial subrepresentation in $\mathrm{Sym}^2(M_2^\theta(\Gamma_g(2)))$ is then spanned by the invariant $\sum_\Delta \theta[\Delta]^8$. It is easy to verify that the dimension of the image of $\mathrm{Sym}^2(M_2^\theta(\Gamma_g(2)))$ is larger than $2^{g-1}(2^g+1) =$

$\dim(\mathbf{1} + \sigma_\theta)$ for $g \geq 2$. As the multiplication map is $Sp(2g)$-equivariant it follows that $\mathrm{Sym}^2(M_2^\theta(\Gamma_g(2))) \subset M_4^\theta(\Gamma_g(2))$ (so if $f_1, \ldots, f_N$ is a basis of $M_2^\theta(\Gamma_g(2))$ then the $f_i f_j$ are linearly independent).

3.8.5. Decomposing representations of $Sp(2g)$

Given a representation of a finite group on a complex vector space, one could determine the value of the character of the representation on each conjugacy class and then use the table of irreducible characters of the group to find the decomposition of the representation. However, it is very time consuming to compute these character values in our examples. Thus we take another approach, which has the additional advantage of identifying explicitly certain subrepresentations.

There is one conjugacy class of $Sp(2g)$ which has only $2^{2g} - 1$ elements, the class of the transvections t_v with $v \in \mathbb{Z}_2^{2g} - \{0\}$, see 3.6.1.. If $\rho : Sp(2g) \to GL(V)$ is a complex representation of $Sp(2g)$, the operator

$$C = C_\rho := \sum_{v \neq 0} \rho(t_v) \qquad (\in GL(V))$$

obviously satisfies $\rho(g)C\rho(g)^{-1} = C$ for all $g \in Sp(2g)$. If $V = \oplus V_i^{n_i}$ is the decomposition of V into irreducible representations V_i, $V_i \not\cong V_j$ if $i \neq j$, then, by Schur's lemma, C must be scalar multiplication by a $\lambda_i \in \mathbb{C}$ on V_i. In particular, the eigenvalues of C are the λ_i with multiplicity $n_i = \dim V_i$ (but it can happen that $\lambda_i = \lambda_j$ for $i \neq j$).

To find λ_i we consider the trace of C on V_i: as the t_v, $v \neq 0$, are the elements of one conjugacy class,

$$Tr(C_{|V_i}) = (2^{2g} - 1)Tr(\rho_i(t_v)) = (2^{2g} - 1)\chi_i(t_v)$$

where t_v is now one specific (but arbitrary) transvection and χ_i is the character of the irreducible representation ρ_i. On the other hand,

$$Tr(C_{|V_i}) = (\dim V_i)\lambda_i, \qquad \text{hence} \quad \lambda_i = \frac{(2^{2g} - 1)\chi_i(t_v)}{\dim V_i}. \tag{60}$$

Note that $\ker(C - \lambda I)$ will be the direct sum of the $V_i^{n_i}$ with $\lambda_i = \lambda$, so we do not only get information on the multiplicities of the irreducible constituents of ρ but also on the corresponding subspaces of V.

These are some of the ingredients we will use to construct string amplitudes and to prove their uniqueness for $g \leq 3$ and in a weaker form for $g = 4$. The explicit construction will clarify the various steps and for $g \leq 2$ we will obtain the known results.

4. Some Remarks on the Constraints

In this section we revise the three constraints for the functions $\Xi_8[\Delta]$ using some mathematical tools introduced in the previous section and we compare them with those of D'Hoker and Phong (DHP) in [DP6] for the functions $\Xi_6[\Delta]$.

- *Remark on condition (ii).* The only essential difference is in the constraint (ii), the transformation request. Note that the products $\theta[\Delta]^4(\tau)\Xi_6[\Delta](\tau)$, with $\Xi_6[\Delta]$ as in the DHP constraint (ii) and $\tau \in \mathbb{H}_3$, transform in the same way as our $\Xi_8[\Delta]$ apart from a factor $\epsilon(M,\Delta)^{4+4}$. However, $\epsilon(M,\Delta)^8 = 1$, so that this difference is only apparent. Conversely, if we require for each $\Xi_8[\Delta]$ to factorize in the product of $\theta[\Delta]^4$ and another function, the latter would satisfy constraint (ii) of [DP6]. But there is not a priory any reason to assume such a factorization: our form for the transformation constraint is weaker than the one imposed in [DP6] because is just a request on the weight of the ' modular forms" and it does not impose a factorized form as a product of a function of weight six times a theta constant at the fourth power. Moreover, it is equivalent to assert that the $\Xi_8[\Delta]$ are modular forms of genus g and weight 8 on $\Gamma_g(2)$.

- *Remark on condition (iii).* DHP impose the factorization condition for an arbitrary separating degeneration. The point is that any such degeneration can be obtained from the one in condition (iii) by a symplectic transformation, so that we have to consider the functions $\Xi_8[\Delta](N \cdot \tau_{k,g-k})$ for all $N \in Sp(2g,\mathbb{Z})$. By constraint (ii), this amounts to considering the function $\Xi_8[N^{-1} \cdot \Delta](\tau_{k,g-k})$ (up to an easy factor) which is indeed determined by constraint (iii).

- *Oversimplification of the problem.* In order to search for a solution of the constraints, we will now show that equation (22) can be used to restrict the problem to a single value of the characteristic, for which we choose $\Delta = \left[\begin{smallmatrix}0\\0\end{smallmatrix}\right]$ with $\left[\begin{smallmatrix}0\\0\end{smallmatrix}\right] = \left[\begin{smallmatrix}0\dots0\\0\dots0\end{smallmatrix}\right]$. In particular, we will see that the problem reduces itself to restrict the constraints to simpler ones for the function $\Xi_8\left[\begin{smallmatrix}0\\0\end{smallmatrix}\right]$. Solving the problem for $\Xi_8\left[\begin{smallmatrix}0\\0\end{smallmatrix}\right]$ will permit us to define functions $\Xi_8[\Delta^{(g)}]$, for all even characteristics $\Delta^{(g)}$, which satisfy the constraints from section 2..

  If we take M to be in the stabilizer of the null characteristic, then condition (ii) is equivalent to require for $\Xi_8\left[\begin{smallmatrix}0\\0\end{smallmatrix}\right]$ to be a modular form on $\Gamma_g(1,2)$ of weight 8. We know that the group $Sp(2g,\mathbb{Z})$ acts transitively on the even characteristics. This means that for any even characteristic $[\Delta^{(g)}]$ there exists at least an $M \in Sp(2g,\mathbb{Z})$ such that $M \cdot \left[\begin{smallmatrix}0\\0\end{smallmatrix}\right] = [\Delta^{(g)}] \bmod 2$. Then we define

$$\Xi_8[\Delta^{(g)}](\tau) := \gamma(M, M^{-1} \cdot \tau)^8 \Xi_8\left[\begin{smallmatrix}0\\0\end{smallmatrix}\right](M^{-1} \cdot \tau), \tag{61}$$

  with $\gamma(M,\tau) := \det(C\tau + D)$. The definition of $\Xi_8[\Delta^{(g)}]$ does not depend on the choice of M and also satisfies the transformation constraint, as shown in [CDG1]. It verifies all the constraints if $\Xi_8\left[\begin{smallmatrix}0\\0\end{smallmatrix}\right]$ satisfies the reduced constraints

(ii$_0$) The function $\Xi_8\left[\begin{smallmatrix}0\\0\end{smallmatrix}\right]$ is a modular form Ξ_8 of weight 8 on $\Gamma_g(1,2)$.

(iii$_0$)(1) For all k, $0 < k < g$, and all $\tau_{k,g-k} \in \Delta_{k,g-k}$ we have

$$\Xi_8\left[\begin{smallmatrix}0\\0\end{smallmatrix}\right](\tau_{k,g-k}) = \Xi_8\left[\begin{smallmatrix}0\\0\end{smallmatrix}\right](\tau_k)\Xi_8\left[\begin{smallmatrix}0\\0\end{smallmatrix}\right](\tau_{g-k})$$

(iii$_0$)(2) If $\Delta^{(g)} = \left[\begin{smallmatrix} ab\dots \\ cd\dots \end{smallmatrix}\right]$ with $ac = 1$ then $\Xi_8[\Delta^{(g)}](\tau_{1,g-1}) = 0$.

(iii$_0$)(1,2) is obviously a consequence of (iii). For (iii$_0$)(1), let us consider the characteristic

$$\Delta^{(g)} = \left[\begin{smallmatrix} a_1\dots a_g \\ b_1\dots b_g \end{smallmatrix}\right], \qquad \Delta^{(k)} := \left[\begin{smallmatrix} a_1\dots a_k \\ b_1\dots b_k \end{smallmatrix}\right], \quad \Delta^{(g-k)} := \left[\begin{smallmatrix} a_{k+1}\dots a_g \\ b_{k+1}\dots b_g \end{smallmatrix}\right].$$

and assume that $\Delta^{(k)}$ is even, so that also $\Delta^{(g-k)}$ is even. By transitivity, there are two symplectic matrices $M_1 \in Sp(2k, \mathbb{Z})$ and $M_2 \in Sp(2(g-k), \mathbb{Z})$ such that $M_1 \cdot \left[\begin{smallmatrix} 0 \\ 0 \end{smallmatrix}\right] = [\Delta^{(k)}]$ and $M_2 \cdot \left[\begin{smallmatrix} 0 \\ 0 \end{smallmatrix}\right] = [\Delta^{(g-k)}]$. We compose such matrices in a block diagonal form so defining a matrix $M \in Sp(2g, \mathbb{Z})$ which has the properties: $M \cdot (\Delta_{k,g-k}) = \Delta_{k,g-k}$ and $\Delta^{(g)} = M \cdot \left[\begin{smallmatrix} 0 \\ 0 \end{smallmatrix}\right]$. As M and $\tau_{k,g-k}$ are made up of $k \times k$ and $(g-k) \times (g-k)$ blocks we get

$$\gamma(M, M^{-1} \cdot \tau_{k,g-k}) = \gamma(M_1, M_1^{-1} \cdot \tau_k)\gamma(M_2, M_2^{-1} \cdot \tau_{g-k}),$$

with $\Delta^{(g)} = M \cdot \left[\begin{smallmatrix} 0 \\ 0 \end{smallmatrix}\right]$. It follows that $M^{-1} \cdot \tau_{k,g-k} \in \Delta_{k,g-k}$ is a matrix with blocks $M_1^{-1} \cdot \tau_k$ and $M_2^{-1} \cdot \tau_{g-k}$. Thus, from the constraint (iii$_0$)(1) get

$$\Xi_8\left[\begin{smallmatrix} 0 \\ 0 \end{smallmatrix}\right](M^{-1} \cdot \tau_{k,g-k}) = \Xi_8\left[\begin{smallmatrix} 0 \\ 0 \end{smallmatrix}\right](M_1^{-1} \cdot \tau_k)\Xi_8\left[\begin{smallmatrix} 0 \\ 0 \end{smallmatrix}\right](M_2^{-1} \cdot \tau_{g-k}),$$

and then we finally have

$$\begin{aligned} \Xi_8[\Delta^{(g)}](\tau_{k,g-k}) &= \gamma(M, M^{-1} \cdot \tau)^8 \Xi_8\left[\begin{smallmatrix} 0 \\ 0 \end{smallmatrix}\right](M^{-1} \cdot \tau_{k,g-k}) \\ &= \gamma(M_1, M_1^{-1} \cdot \tau_k)^8 \gamma(M_2, M_2^{-1} \cdot \tau_{g-k})^8 \\ &\quad \cdot \Xi_8\left[\begin{smallmatrix} 0 \\ 0 \end{smallmatrix}\right](M_1^{-1} \cdot \tau_k)\Xi_8\left[\begin{smallmatrix} 0 \\ 0 \end{smallmatrix}\right](M_2^{-1} \cdot \tau_{g-k}) \\ &= \Xi_8[\Delta^{(k)}](\tau_k)\Xi_8[\Delta^{(g-k)}](\tau_{g-k}), \end{aligned}$$

so for such $\Delta^{(g)}$ the functions $\Xi_8[\Delta^{(g)}]$ satisfy (iii).

5. Existence and Uniqueness of the Forms $\Xi_8[\Delta]$

We are now ready to work out a solution for the constraints and show its uniqueness, at least for genus $g \leq 3$. For genus four the uniqueness will be proved in a weakened form that is by restricting to polynomial expressions. We will analyze the equations genus by genus.

5.1. The case $g = 1$

In genus one there are three even characteristics and one odd. Using the classical theta formula (54), we can find the relations between the classical and the second order theta constants. Moreover, there are no algebraic relations between the $\Theta[\sigma]$'s. The dimension formula (??) and the fact that $M_{2k}^{\theta}(\Gamma_1(2)) = M_{2k}(\Gamma_1(2))$ show that $\dim M_{2k}(\Gamma_1(2)) = k + 1$. A basis of $M_2(\Gamma_1(2))$ is given by the Heisenberg invariants $\Theta[0]^4 + \Theta[1]^4$ and

$(\Theta[0]\Theta[1])^2$ and a basis of $M_{2k}(\Gamma_1(2)$ is given by homogeneous polynomials of degree k in these invariants:

$$(\Theta[0]^4 + \Theta[1]^4)^k, \quad (\Theta[0]\Theta[1])^2(\Theta[0]^4 + \Theta[1]^4)^{k-1}, \quad \ldots, \quad (\Theta[0]\Theta[1])^{2k}.$$

In genus one the group $\mathrm{Sp}(2,\mathbb{Z})$ is isomorphic to the special linear group $SL(2,\mathbb{Z})$ and its standard generators are S and T. The classical transformation theory of theta functions gives:

$$\rho_2(S): \begin{cases} \theta\left[\begin{smallmatrix}0\\0\end{smallmatrix}\right]^4 & \longmapsto & -\theta\left[\begin{smallmatrix}0\\0\end{smallmatrix}\right]^4 \\ \theta\left[\begin{smallmatrix}0\\1\end{smallmatrix}\right]^4 & \longmapsto & -\theta\left[\begin{smallmatrix}1\\0\end{smallmatrix}\right]^4 \\ \theta\left[\begin{smallmatrix}1\\0\end{smallmatrix}\right]^4 & \longmapsto & -\theta\left[\begin{smallmatrix}0\\1\end{smallmatrix}\right]^4 \end{cases}, \qquad \rho_2(T): \begin{cases} \theta\left[\begin{smallmatrix}0\\0\end{smallmatrix}\right]^4 & \longmapsto & \theta\left[\begin{smallmatrix}0\\1\end{smallmatrix}\right]^4 \\ \theta\left[\begin{smallmatrix}0\\1\end{smallmatrix}\right]^4 & \longmapsto & \theta\left[\begin{smallmatrix}0\\0\end{smallmatrix}\right]^4 \\ \theta\left[\begin{smallmatrix}1\\0\end{smallmatrix}\right]^4 & \longmapsto & -\theta\left[\begin{smallmatrix}1\\0\end{smallmatrix}\right]^4 \end{cases}.$$

Here, $S = \left(\begin{smallmatrix}0&1\\-1&0\end{smallmatrix}\right)$ and $T = \left(\begin{smallmatrix}1&1\\0&1\end{smallmatrix}\right)$ are the standard generators of $SL(2,\mathbb{Z})$. In the computation of the matrices of the $\rho_2(g)$'s w.r.t. the basis of $M_2(\Gamma_1(2))$ ($\theta\left[\begin{smallmatrix}0\\0\end{smallmatrix}\right]$ and $\theta\left[\begin{smallmatrix}0\\1\end{smallmatrix}\right]$, for example) one has to apply the Jacobi relation $\theta\left[\begin{smallmatrix}0\\0\end{smallmatrix}\right]^4 = \theta\left[\begin{smallmatrix}1\\0\end{smallmatrix}\right]^4 + \theta\left[\begin{smallmatrix}0\\1\end{smallmatrix}\right]^4$. As follows from section 3.8.4. and from the table of characters of S_3 reported in Table 5, $M_2(\Gamma_1(2))$ is the unique irreducible two dimensional representation of the symmetric group S_3 that can be identified with the representation $\rho_\theta = \rho[21]$, hence $M_{2k}(\Gamma_1(2)) \simeq \mathrm{Sym}^k(\rho[21])$. The group $O^+(2)$ is the

Table 5. Table of the characters of the group S_3.

S_3	$C_{1,1,1}$	$C_{2,1}$	C_3
$\rho_{[1^3]}$	1	1	1
$\rho_{[2,1]}$	1	−1	1
$\rho_{[3]}$	2	0	−1

group of order two generated by the image of $S \in SL(2,\mathbb{Z})$ in $Sp(2)$.

We now study to the decomposition of $M_6(\Gamma_1(2))$ because some functions we will extensively use in the construction of superstring measures belong to this space. The representation on $M_6(\Gamma_1(2)) = \mathrm{Sym}^3(M_2(\Gamma_1(2)))$ can be decomposed in irreducible representations of S_3 as:

$$M_6(\Gamma_1(2)) \cong \rho_{[3]} \oplus \rho_{[2,1]} \oplus \rho_{[1^3]},$$

where $\rho_{[1^3]}$ is the sign representation of S_3 and $\rho_{[3]}$ is the trivial representation. One obtains this decomposition using the techniques we will expose for the more enlightening case $g = 2$, see section 5.2.1..

It is easy to verify that the three irreducible representations are generated by:

$$\begin{aligned} \rho_{[3]} &= \langle \Theta[0]^{12} - 33\Theta[0]^8\Theta[1]^4 - 33\Theta[0]^4\Theta[1]^8 + \Theta[1]^{12} \rangle, \\ \rho_{[1^3]} &= \langle \eta^{12} \rangle, \\ \rho_{[2,1]} &= \{(a+b)\theta\left[\begin{smallmatrix}0\\0\end{smallmatrix}\right]^{12} + a\theta\left[\begin{smallmatrix}0\\1\end{smallmatrix}\right]^{12} + b\theta\left[\begin{smallmatrix}1\\0\end{smallmatrix}\right]^{12} : a, b \in \mathbb{C}\}, \end{aligned}$$

where we used a classical formula for the Dedekind η function: $\eta^3 = \theta\begin{bmatrix}0\\0\end{bmatrix}\theta\begin{bmatrix}0\\1\end{bmatrix}\theta\begin{bmatrix}1\\0\end{bmatrix}$, so

$$\begin{aligned}\eta^{12} &= \theta\begin{bmatrix}0\\0\end{bmatrix}^4\theta\begin{bmatrix}0\\1\end{bmatrix}^4\theta\begin{bmatrix}1\\0\end{bmatrix}^4\\ &= (\Theta[0]^2+\Theta[1]^2)^2(\Theta[0]^2-\Theta[1]^2)^2(2\Theta[0]\Theta[1])^2.\end{aligned}$$

The two dimensional subspace of $O^+(2)$-anti-invariants, see section 3.8.3., is:

$$M_6(\Gamma_g(2))^\epsilon = \langle \eta^{12},\ f_{21} := 2\theta\begin{bmatrix}0\\0\end{bmatrix}^{12} + \theta\begin{bmatrix}0\\1\end{bmatrix}^{12} + \theta\begin{bmatrix}1\\0\end{bmatrix}^{12}\rangle,$$

the function f_{21} lies in the two-dimensional $\rho_{[21]}$ irreducible subrepresentation[13]. We list here some other modular forms that can be expressed in terms of f_{21} and η^{12} that we will use later.

$$\begin{aligned}\theta^{12}\begin{bmatrix}0\\0\end{bmatrix} &= \tfrac{1}{3}f_{21}+\eta^{12},\\ \theta\begin{bmatrix}0\\0\end{bmatrix}^4(\theta\begin{bmatrix}0\\0\end{bmatrix}^8+\theta\begin{bmatrix}0\\1\end{bmatrix}^8+\theta\begin{bmatrix}1\\0\end{bmatrix}^8) &= \tfrac{2}{3}f_{21},\\ \theta\begin{bmatrix}0\\0\end{bmatrix}^{12}+\theta\begin{bmatrix}0\\1\end{bmatrix}^{12}+\theta\begin{bmatrix}1\\0\end{bmatrix}^{12} &= \tfrac{2}{3}f_{21}-\eta^{12},\\ \theta\begin{bmatrix}0\\0\end{bmatrix}^4\theta\begin{bmatrix}0\\1\end{bmatrix}^8+\theta\begin{bmatrix}0\\0\end{bmatrix}^4\theta\begin{bmatrix}1\\0\end{bmatrix}^8 &= \tfrac{1}{3}f_{21}-\eta^{12}.\end{aligned}$$

In genus one it is well known that the modular forms $\Xi_8[\Delta]$ are given by $\Xi_8[\Delta] = \theta[\Delta]^4\eta^{12}$. The function $\Xi_8\begin{bmatrix}0\\0\end{bmatrix} = \theta\begin{bmatrix}0\\0\end{bmatrix}^4\eta^{12}$ is a modular form of weight eight on $\Gamma_1(1,2)$.

5.2. The case $g = 2$

At genus two there are ten even characteristics which correspond to ten theta constants. From the Table 2 we know that the ring of the Heisenberg invariants of degree four is a five dimensional space. It is generated by the fourth power of the theta constants, as can be shown using the classical formula (and a computer to make the computations faster). A useful basis for this space, obtained as explained in section 3.6., is provided by the following five homogeneous polynomials $p_0, \cdots, p_4$ of degree four in the $\Theta[\sigma]$'s:

$$\mathbb{C}[\ldots,\Theta[\sigma],\ldots]^{H_2} = \mathbb{C}[p_0,\ldots,p_4] \tag{62}$$

where the p_i are defined by

$$\begin{aligned}p_0 &= \Theta[00]^4+\Theta[01]^4+\Theta[10]^4+\Theta[11]^4,\\ p_1 &= 2(\Theta[00]^2\Theta[01]^2+\Theta[10]^2\Theta[11]^2),\\ p_2 &= 2(\Theta[00]^2\Theta[10]^2+\Theta[01]^2\Theta[11]^2),\\ p_3 &= 2(\Theta[00]^2\Theta[11]^2+\Theta[01]^2\Theta[10]^2),\\ p_4 &= 4\Theta[00]\Theta[01]\Theta[10]\Theta[11].\end{aligned} \tag{63}$$

By means of the classical formula we can expand the ten $\theta[\delta]^4$ on this basis. We summarize the result in Table 6, for example:

$$\theta[\delta_7]^4 = 2p_2 + 2p_4. \tag{64}$$

Table 6. Expansion of $\theta^4[\delta]$ on the basis of p_i

δ	$\theta^4[\delta]$	p_0	p_1	p_2	p_3	p_4
δ_1	$\theta^4\left[\begin{smallmatrix}0&0\\0&0\end{smallmatrix}\right]$	1	1	1	1	0
δ_2	$\theta^4\left[\begin{smallmatrix}0&0\\0&1\end{smallmatrix}\right]$	1	-1	1	-1	0
δ_3	$\theta^4\left[\begin{smallmatrix}0&0\\1&0\end{smallmatrix}\right]$	1	1	-1	-1	0
δ_4	$\theta^4\left[\begin{smallmatrix}0&0\\1&1\end{smallmatrix}\right]$	1	-1	-1	1	0
δ_5	$\theta^4\left[\begin{smallmatrix}0&1\\0&0\end{smallmatrix}\right]$	0	2	0	0	2
δ_6	$\theta^4\left[\begin{smallmatrix}0&1\\1&0\end{smallmatrix}\right]$	0	2	0	0	-2
δ_7	$\theta^4\left[\begin{smallmatrix}1&0\\0&0\end{smallmatrix}\right]$	0	0	2	0	2
δ_8	$\theta^4\left[\begin{smallmatrix}1&0\\0&1\end{smallmatrix}\right]$	0	0	2	0	-2
δ_9	$\theta^4\left[\begin{smallmatrix}1&1\\0&0\end{smallmatrix}\right]$	0	0	0	2	2
δ_{10}	$\theta^4\left[\begin{smallmatrix}1&1\\1&1\end{smallmatrix}\right]$	0	0	0	2	-2

The advantage of the selected basis is that it defines a map $\mathbb{P}^3 \to \mathbb{P}^4$, $(\Theta[00](\tau) : \Theta[01](\tau) : \Theta[10](\tau) : \Theta[11](\tau)) \to (p_0 : p_1 : p_2 : p_3 : p_4)$ and the image of $\mathbb{P}^3$ results to be defined by a quartic polynomial f_4 in 5 variables, the Igusa quartic, so that $I_4 = f_4(p_0, p_1, p_2, p_3, p_4)$ is identically zero. Its explicit expression is

$$I_4 = p_4^4 + p_4^2 p_0^2 - p_4^2 p_1^2 - p_4^2 p_2^2 - p_4^2 p_3^2 + p_1^2 p_2^2 + p_1^2 p_3^2 + p_2^2 p_3^2 - 2p_0 p_1 p_2 p_3. \tag{65}$$

Expressing the p_i in term of the four second order theta constant one verifies that this polynomial vanishes. We can also write I_4 using the ten classical theta constants obtaining:

$$I_4 = \frac{1}{192}\left[\left(\sum_\delta \theta^8[\delta]\right)^2 - 4\sum_\delta \theta^{16}[\delta]\right]. \tag{66}$$

It thus provides a relation of order 16 in the classical $\theta[\delta]$. From this we see that, as a graded ring,

$$\oplus_{k=0}^{\infty} M_{2k}(\Gamma_2(2)) \cong \mathbb{C}[y_0, \ldots, y_4]/(f_4(y_0, \ldots, y_4)), \tag{67}$$

and for $k \leq 3$ there are not constraints, so that the dimensions are

$$\dim M_2(\Gamma_2(2)) = 5, \qquad \dim M_4(\Gamma_2(2)) = \binom{5+1}{2} = 15,$$

$$\dim M_6(\Gamma_2(2)) = \binom{5+2}{3} = 35, \qquad \dim M_8(\Gamma_2(2)) = \binom{5+3}{4} - 1 = 69.$$

[13] From this we choose the "strange" name f_{21}.

5.2.1. The Sp(4)-representations on the $M_{2k}(\Gamma_2(2))$

We will now study the representations of the symplectic group on the space of modular forms. We use the isomorphism $\mathrm{Sp}(4) \cong S_6$ and the irreducible representations will be labeled by partitions of 6. Table 7 collects the characters of the group S_6. In the second column are given the partitions of 6 and in the first one the names we have chosen for the corresponding representations. The symmetric group S_6 has eleven conjugacy classes

Table 7. Characters of the conjugacy classes of the eleven irreducible representations of S_6.

S_6	Partition	C_1	C_2	C_3	$C_{2,2}$	C_4	$C_{3,2}$	C_5	$C_{2,2,2}$	$C_{3,3}$	$C_{4,2}$	C_6
id_1	$[6]$	1	1	1	1	1	1	1	1	1	1	1
alt_1	$[1^6]$	1	-1	1	1	-1	-1	1	-1	1	1	-1
st_5	$[2^3]$	5	-1	-1	1	1	-1	0	3	2	-1	0
sta_5	$[3^2]$	5	1	-1	1	-1	1	0	-3	2	-1	0
rep_5	$[5\,1]$	5	3	2	1	1	0	0	-1	-1	-1	-1
repa_5	$[2\,1^4]$	5	-3	2	1	-1	0	0	1	-1	-1	1
n_9	$[4\,2]$	9	3	0	1	-1	0	-1	3	0	1	0
na_9	$[2^2\,1^2]$	9	-3	0	1	1	0	-1	-3	0	1	0
sw_{10}	$[3\,1^3]$	10	-2	1	-2	0	1	0	2	1	0	-1
swa_{10}	$[4\,1^2]$	10	2	1	-2	0	-1	0	-2	1	0	1
s_{16}	$[3\,2\,1]$	16	0	-2	0	0	0	1	0	-2	0	0

so that it has eleven irreducible representations. For example, the class $C_{3,2}$ consists of the product of a two-cycle and a three-cycle, and the character of the first ten dimensional representation, sw_{10} or $[3\,1^3]$ in standard notation, for this class is 1.

To find the correspondence among the symplectic group and the symmetric group we can relate the transformations of the theta constants under the action of the generators of $\mathrm{Sp}(4)$, given by the transformation formula for the theta constants (41), to the transformations of the same functions obtained by the action of S_6 on the branch points appearing in the Thomae formula. The generators of $\mathrm{Sp}(4)$ are:

$$
\begin{aligned}
&M_i = \begin{pmatrix} I & B_i \\ 0 & I \end{pmatrix}, \quad B_1 = \begin{pmatrix} 1 & 0 \\ 0 & 0 \end{pmatrix}, \quad B_2 = \begin{pmatrix} 0 & 0 \\ 0 & 1 \end{pmatrix}, \\
&B_3 = \begin{pmatrix} 0 & 1 \\ 1 & 0 \end{pmatrix}; \\
&S = \begin{pmatrix} 0 & I \\ -I & 0 \end{pmatrix}; \quad \Sigma = \begin{pmatrix} \sigma & 0 \\ 0 & -\sigma \end{pmatrix}, \quad \sigma = \begin{pmatrix} 0 & 1 \\ -1 & 0 \end{pmatrix}; \\
&T = \begin{pmatrix} \tau_+ & 0 \\ 0 & \tau_- \end{pmatrix}, \quad \tau_+ = \begin{pmatrix} 1 & 1 \\ 0 & 1 \end{pmatrix}, \quad \tau_- = \begin{pmatrix} 1 & 0 \\ -1 & 1 \end{pmatrix}.
\end{aligned}
\tag{68}
$$

The phase factor $\epsilon(\delta, M)$, satisfying $\epsilon^8(\delta, M) = 1$, depends both on the characteristic δ and on the matrix M generating the transformation[14]. For the even characteristics $\delta = \left[\begin{smallmatrix} a \\ b \end{smallmatrix}\right]$ the fourth powers of ϵ are given by:

$$\begin{aligned} \epsilon^4(\delta, M_i) &= e^{\pi i\, {}^t a B_i a} \qquad i = 1, 2, \\ \epsilon^4(\delta, M_3) &= \epsilon^4(\delta, S) = \epsilon^4(\delta, \Sigma) = \epsilon^4(\delta, T) = 1. \end{aligned} \tag{69}$$

In Table 8 we report the relationship between the generators of the modular group and S_6.

Table 8. Relationship between the generators of the modular group and S_6.

M_1	M_2	M_3	S	Σ	T
$(1\,3)$	$(2\,4)$	$(1\,3)(2\,4)(5\,6)$	$(3\,5)(4\,6)$	$(1\,2)(3\,4)(5\,6)$	$(1\,3)(2\,6)(4\,5)$

We now need to identify the representation $M_2(\Gamma_2(2))$. We know from the result of van Geemen in section 3.8.4. that this five dimensional representation is ρ_θ which is irreducible. The characters table of S_6 shows that there are four five dimensional irreducible representations. To find which one is supported by the $\theta[\delta]^4$'s we study the action of the permutation (12), that belongs to the conjugacy class C_2, on the basis $\theta[\delta_i]$, $i = 1, \cdots, 2$ of $M_2(\Gamma_2(2))$ (alternatively one can study the action of one generator of $\mathrm{Sp}(4)$, for example M_1, obtaining the same result). The matrix associated to this permutation is:

$$M_{1\,2} = \begin{pmatrix} 1 & 0 & 0 & 0 & 0 \\ -1 & -1 & 0 & 0 & 0 \\ 0 & 0 & -1 & -1 & 0 \\ 0 & 0 & 0 & 1 & 0 \\ -1 & 0 & 0 & 1 & -1 \end{pmatrix}. \tag{70}$$

The trace of this matrix is -1 and this is exactly the character of the representation spanned by the $\theta[\delta]^4$'s. Thus $\rho_\theta = \rho_{[2^3]}$, i.e. the representation is the st_5. More in general, using the theory of representations of S_6 we find:

$$\begin{aligned} M_2(\Gamma_2(2)) &\cong \rho_{[2^3]}, \\ M_4(\Gamma_2(2)) &\cong \mathrm{Sym}^2(\rho_{[2^3]}) &&= \mathbf{1} + \rho_{[42]} + \rho_{[2^3]}, \\ M_6(\Gamma_2(2)) &\cong \mathrm{Sym}^3(\rho_{[2^3]}) &&= \mathbf{1} + 2\rho_{[2^3]} + \rho_{[21^4]} + \rho_{[42]} + \rho_{[31^3]}, \\ M_8(\Gamma_2(2)) &\cong \mathrm{Sym}^4(\rho_{[2^3]}) - \mathbf{1} &&= \mathbf{1} + 3\rho_{[2^3]} + 3\rho_{[42]} + \rho_{[31^3]} + \rho_{[321]}. \end{aligned}$$

From these decompositions we can conclude that $\rho_r = \rho_{[21^4]}$ and $\sigma_\theta = \rho_{[42]}$ (from the next discussion or cf. [CD] one can also deduce that $\rho_\theta + \rho_r$ is the representation on the $\theta[\delta]^{12}$, which is $\mathrm{Ind}_{O^+}^{Sp}(\epsilon)$ and $1 + \sigma_\theta$ is the permutation representation on the 10 even $\theta[\delta]^8$, hence its trace must be ≥ 0 on each conjugacy class). The computation of the symmetric powers of

[14] In ϵ there are both the contributes of $\kappa(M)$ and of $\Phi_\Delta(M)$ of (41)

a representation can be done using a computer (for example with Magma or Mathematica), but also by hand. We expose the details for the computation of $\mathrm{Sym}^3(\rho_{[2^3]})$, the others being similar. To establish the characters of the representations on $\mathrm{Sym}^3(M_2(\Gamma_2(2)))$ we proceed as follows. If $\mathrm{Tr}(\rho^{V_\theta}(g)) = \sum_i \lambda_i$, where to shorten the notations we posed $V_\theta \equiv M_2(\Gamma_2(2))$, then for $g \in S_6$:

$$\mathrm{Tr}(\rho^{V_\theta}(g^2)) = \sum_i \lambda_i^2, \qquad \mathrm{Tr}(\rho^{V_\theta}(g^3)) = \sum_i \lambda_i^3, \tag{71}$$

$$(\mathrm{Tr}(\rho^{V_\theta}))^3 = \sum_i \lambda_i^3 + 3\sum_{i\neq j} \lambda_i^2\lambda_j + 6 \sum_{1\leq i<j<k\leq 5} \lambda_i\lambda_j\lambda_k,$$

$$\mathrm{Tr}(\rho^{S^3V_\theta}(g)) = \sum_{1\leq i\leq j\leq k\leq 5} \lambda_i\lambda_j\lambda_k.$$

Using the previous relations we finally obtain

$$\begin{aligned}\mathrm{Tr}(\rho^{S^3V_\theta}(g)) &= \frac{1}{6}(\mathrm{Tr}(\rho^{V_\theta}(g)))^3 + \frac{1}{2}\mathrm{Tr}(\rho^{V_\theta}(g))\mathrm{Tr}(\rho^{V_\theta}(g^2)) \\ &+ \frac{1}{3}\mathrm{Tr}(\rho^{V_\theta}(g^3)).\end{aligned} \tag{72}$$

To apply this formula we have to know in which conjugacy class are g^2 and g^3. For example, for the element $g = (12)$ its square is in the class of the identity, C_1, and its cube in the same class of g, C_2. Hence, we obtain that the character of the representation of $g = (12)$ on $\mathrm{Sym}^3 V_\theta = \mathrm{Sym}^3(M_2(\Gamma_2(2)))$ is -3. In the same way we can compute the characters χ of the other ten conjugacy classes:

$$\chi(\rho_{V_\theta}) = \{35, -3, 2, 3, 1, 0, 0, 13, 5, -1, 1\} \tag{73}$$

The representation on $\mathrm{Sym}^3 V_\theta$ will be a direct sum of the irreducible representations of S_6: $\mathrm{Sym}^3 V_\theta = \oplus_i m_i \rho_i$. Using the orthogonality of the characters we can compute the m_i as $m_i = \langle \chi_{\mathrm{Sym}^3 V_\theta}, \chi_i\rangle$. The inner product between characters is given by $\langle \chi, \psi\rangle = \frac{1}{|S_6|}\sum_{g\in S_6}\chi(g)\overline{\psi(g)} = {}^t\chi T\overline{\psi}$, where T is the matrix

$$\begin{aligned} T &= \frac{1}{|S_6|}\,\mathrm{diag}\{|C_1|, \cdots, |C_6|\} \\ &= \frac{1}{720}\,\mathrm{diag}\{1, 15, 40, 45, 90, 120, 144, 15, 40, 90, 120\}.\end{aligned} \tag{74}$$

On the diagonal there are the numbers of elements in each of the eleven conjugacy classes of S_6. Then, we get the complete decomposition

$$\mathrm{Sym}^3(\mathrm{st}_5) = \mathrm{id}_1 + \mathrm{n}_9 + \mathrm{repa}_5 + 2\mathrm{st}_5 + \mathrm{sw}_{10}, \tag{75}$$

which expresses exactly the desired result.
In the next subsection, in order to solve the constraints, we will construct the basis for the

space of modular forms of weight eight in which there is the representations $\mathbf{1}$ and σ_θ, as well the space of modular form of weight 6 in which there is the representation ρ_θ. Here we conclude our analysis by looking at which representation is supported by some particular vector spaces. We will focus on the space $\mathrm{Sym}^3(M_2(\Gamma_2(2)))$, clarifying the origin of the functions Ξ_6 introduced by D'Hoker and Phong, and on some other spaces generated by polynomial of degree 12 in the theta constants.

Let us consider the matrix representation of S_6 in the space $\mathrm{Sym}^3 V_\theta$ and define the matrix $N_2 \in \mathrm{Mat}(35, \mathbb{Z})$ as the sum of the 15 matrices representing the elements in the conjugacy class C_2. If $N(\sigma)$ is the matrix representing any element $\sigma \in S_6$, clearly we have $N(\sigma)N_2N(\sigma)^{-1} = N_2$, as $N(\sigma)$ just changes the order of the addends in the sum. The eigenvalues of this matrix are

$$N_{2\,\mathrm{diag}} = \mathrm{diag}\{15, \underbrace{-9, \cdots, -9}_{5 \text{ times}}, \underbrace{5, \cdots, 5}_{9 \text{ times}}, \underbrace{-3 \cdots, -3}_{20 \text{ times}}\}. \tag{76}$$

and its trace is -45. From Schur lemma, the matrix N_2 acts as a multiple of the identity over each subspace V_i supporting an irreducible representation, i.e. $N_2|_{V_i} = \lambda_i^{(2)} \,\mathrm{Id}\,|_{V_i}$ and from (60):

$$\lambda_i^{(2)} = \frac{15\mathrm{Tr}\rho_i(1\,2)}{\dim(V_i)}, \tag{77}$$

where we chosen $g = (1\,2)$ as two cocycle of S_6. We can deduce many interesting informations from this computation, even though this matrix is not enough to decompose the whole space $\mathrm{Sym}^3 V_\theta$: indeed, from (77) we deduce that the eigenvalue 15 corresponds to the representation id_1 of dimension one, the eigenvalue -9 to representation repa_5 of dimension five and the eigenvalue -5 to the representation n_9 of dimension nine. But for the last twenty eigenvalues we are not able to distinguish the spaces with the same eigenvalue (st_5 and sw_{10}). To recognize univocally all the subspace of $\mathrm{Sym}^3 V_\theta$ we then consider also the conjugacy class $C_{2,2,2}$ of the product of three 2-ciclyes. This class has 15 elements. As before we compute the matrix $N_{2,2,2} \in \mathrm{Mat}(35, \mathbb{Z})$, which for the Schur lemma acts as a multiple of identity on each subspace supporting an irreducible representation, and whose eignevalues are:

$$N_{2,2,2\,\mathrm{diag}} = \mathrm{diag}\{15, \underbrace{9, \cdots 9}_{10 \text{ times}}, \underbrace{5, \cdots, 5}_{9 \text{ times}}, \underbrace{3 \cdots, 3}_{15 \text{ times}}\}. \tag{78}$$

In this case we find that the representation st_5 of dimension five has eigenvalue 9, the representation repa_5 of dimension five has eigenvalue 3 and the representation sw_{10} of dimension ten has eigenvalue 3. We collect all these informations in the Table 9. Note that for each representation there is a different (ordered) couple of eigenvalues. Let W be a (modular invariant) space generated by some degree twelve polynomials in theta constants. Expanding all the polynomials generating W on a basis of $\mathrm{Sym}^3 V_\theta$ we obtains the coefficient matrix M_W with 35 rows and n columns, with $n = \dim W$. Thus, if W coincides with one of the V_i we get

$${}^t(N_C - \lambda_i^C \,\mathrm{Id}) M_W = 0, \tag{79}$$

Table 9. Eigenvalues and dimensions of the eigenspaces appearing in the decomposition of $\mathrm{Sym}^3 V_\theta$**.**

	$\lambda_i^{(2)}$	$\lambda_i^{(2,2,2)}$	$\dim(V_i)$
id_1	15	15	1
st_5	-3	9	5
repa_5	-9	3	5
n_9	5	5	9
sw_{10}	-3	3	10

where N_C stands for N_2 or $N_{2,2,2}$ and λ_i^C the corresponding eigenvalue[15].
The eigenvalue 15 with multiplicity one shows the presence of a subspace V_I of dimension one invariant under the action of S_6, i.e. there is an invariant cubic polynomial in p_i which we will call ψ_6. To find it, we can compute the kernel of the matrix ${}^tN - 15\,\mathrm{Id}_{35}$ (which is evidently one dimensional). We get

$$\Psi_6 = p_0^3 - 9p_0(p_1^2 + p_2^2 + p_3^2 - 4p_4^2) + 54p_1p_2p_3. \tag{80}$$

Using the classical formula we see that this polynomial coincides, up a multiplicative constant, with the modular form of weight six appearing in [DP4]. Let us now consider the space $V_\Xi = \langle \cdots, \Xi_6[\delta], \cdots \rangle$ of the forms introduced by D'Hoker and Phong in [DP4] to define the superstring measures in genus two. We find

$${}^t(N_2 - (-3)\,\mathrm{Id})M_{V_\Xi} = 0, \tag{81}$$
$${}^t(N_{2,2,2} - 9\,\mathrm{Id})M_{V_\Xi} = 0, \tag{82}$$

which shows that V_Ξ supports the representation st_5. The five dimensional subspace $V_S = \langle \theta^4[\delta]\sum_{\delta'} \theta^8[\delta'] \rangle$ provides the second representation st_5:

$${}^t(N_{2,2,2} - 9\,\mathrm{Id})M_S = 0, \quad V_S \neq V_\Xi. \tag{83}$$

align Next, there is the space V_f supporting the representation repa_5, which is given by combinations of the functions $\Xi_6[\delta]$ and the derivatives of the Igusa quartic w.r.t. the $\theta[\delta]^4$, $V_f = \langle \cdots, 2\Xi_6[\delta] - \frac{\partial I_4}{\partial\theta[\delta]^4} \cdots \rangle$:

$${}^t(N_2 - (-9)\,\mathrm{Id})M_{V_f} = 0. \tag{84}$$

To find an expression for the representation spaces n_9 and sw_{10} we proceed as follows. Let us consider a space W (a priori it can be also reducible), compute the matrix M_W and define the product

$$\prod_{i,j} {}^t(N_2 - \lambda_i^{(2)}\,\mathrm{Id}_{35})\,{}^t(N_{2,2,2} - \lambda_j^{(2,2,2)}\,\mathrm{Id}_{35}). \tag{85}$$

[15]The transpose appears because N_C transforms the basis of $\mathrm{Sym}^3 V_theta$ and M_W is the matrix of the coefficients.

The kernel of each factor is the space in which $N^C = N_2$ or $N_{2,2,2}$ has eigenvalue λ_i^C. If a vector does not belong to any representation for which the λ_i appears in the product, its image will be different from the null vector. If in the product (85) we omit a particular representation, say the k-th one, then the expression

$$\prod_{i,j} {}^t(N_2 - \lambda_i \,\mathrm{Id}_{35})\, {}^t(N_{2,2,2} - \lambda_j \,\mathrm{Id}_{35}) M_W \tag{86}$$

is non vanishing if and only if the representation k belong to W. This follows from the fact that on a suitable basis the expression (85) assumes the diagonal form

$$\mathrm{diag}\{0, \cdots, 0, \mu_k, \cdots, \mu_k, 0, \cdots, 0\}, \tag{87}$$

$$\mu_k = \prod_i (\lambda_k^{(2)} - \lambda_i^{(2)}) \prod_j (\lambda_k^{(2,2,2)} - \lambda_j^{(2,2,2)}), \tag{88}$$

giving zero when multiplied by a vector belonging to any representation different from V_k. The multiplicity of μ_k is, clearly, equal to $n_k \dim(V_k)$, where $\mathrm{Sym}^3 = \oplus_k V_k^{n_k}$ and the n_k are computed as the coefficients appearing in the decomposition (75). The subspace W_k giving the representation k will be generated by the image of the product $\prod_{i,j} {}^t(N_2 - \lambda_i \,\mathrm{Id}_{35})\, {}^t(N_{2,2,2} - \lambda_j \,\mathrm{Id}_{35})$ acting on a basis of the whole $\mathrm{Sym}^3 V_\theta$ and one can extract a basis from it. Thus we can decompose the space $\mathrm{Sym}^3 V_\theta$ as:

$$S^3 V_\theta = V_I \oplus V_\Xi \oplus V_f \oplus V_S \oplus V_9 \oplus V_{10}, \tag{89}$$

where the spaces V_9 and V_{10} are suitable spaces of dimension nine and ten respectively and constructed as explained before. Applying the previous approach to some space of polynomials of degree twelve in theta constants we obtain the results reported in Table 5.2.1..

5.2.2. Construction of Ξ_8 at genus 2

We are now able to construct the superstring measure at $g = 2$ and prove its uniqueness. We recall, that $\dim M_8(\Gamma_2(2))^{O^+} = n_1 + n_\theta$, with n_1, n_θ the multiplicity of $\mathbf{1}$ and $\sigma_\theta = \rho_{[42]}$ in $M_8(\Gamma_2(2))$ respectively, see section 3.8.3.. From the decomposition given there we get

$$\dim M_8(\Gamma_2(2))^{O^+} = 1 + 3 = 4.$$

The subspace $M_8(\Gamma_2(2))^{O^+}$ contains the $Sp(4)$-invariant $\sum_\delta \theta[\delta]^{16} \in M_8(\Gamma_2(2))$ as well as the three dimensional subspace spanned by

$$F_1^{(2)} := \theta\left[\begin{smallmatrix}00\\00\end{smallmatrix}\right]^{16}, \quad F_2^{(2)} := \theta\left[\begin{smallmatrix}00\\00\end{smallmatrix}\right]^{4} \sum_\delta \theta[\delta]^{12}, \quad F_3^{(2)} := \theta\left[\begin{smallmatrix}00\\00\end{smallmatrix}\right]^{8} \sum_\delta \theta[\delta]^{8}.$$

The apex indicates the genus and when it is clear from the context we will omit it. Using the classical theta formula, one easily sees that these four functions are linearly independent and thus are a basis of $M_8(\Gamma_2(2))^{O^+}$. The function $\Xi_8[0^{(2)}](\tau_2)$, to satisfy the third constraint, should restrict to $\Xi_8[0^{(1)}](\tau_1)\Xi_8[0^{(1)}](\tau_2)$ with $\Xi_8[0^{(1)}](\tau_1) = (\theta\left[\begin{smallmatrix}0\\0\end{smallmatrix}\right]^4 \eta^{12})(\tau_1)$ on $\mathbb{H}_1 \times \mathbb{H}_1 \subset$

Table 10. Decomposition of some vectorial spaces. We intend: $\partial_{\delta_i} \equiv \frac{\partial}{\partial\theta[\delta_i]^4}$.

Space	Dimension	Representations
$\langle \partial_{\delta_i} I_4 \rangle$	10	$\mathrm{st}_5 \oplus \mathrm{repa}_5$
$\langle \Xi_6[\delta_i] \rangle$	5	st_5
$\langle \theta^{12}[\delta_i] \rangle$	10	$\mathrm{st}_5 \oplus \mathrm{repa}_5$
$\langle \theta^4[\delta_i] \sum_{\delta'} \theta^8[\delta'] \rangle$	5	st_5
$\langle \theta^4[\delta_i] \theta^8[\delta_j] \rangle$	34	$2\mathrm{st}_5 \oplus \mathrm{repa}_5 \oplus \mathrm{n}_9 \oplus \mathrm{sw}_{10}$
$\langle \theta^{12}[\delta_i], \partial_{\delta_j} I_4 \rangle$	15	$\mathrm{st}_5 \oplus \mathrm{st}_5 \oplus \mathrm{repa}_5$
$\langle \theta^{12}[\delta_i], \Xi_6 \rangle$	15	$\mathrm{st}_5 \oplus \mathrm{repa}_5$
$\langle \theta^{12}[\delta_i], \theta^4[\delta_j] \sum_{\delta'} \theta^8[\delta'] \rangle$	15	$\mathrm{st}_5 \oplus \mathrm{repa}_5$
$\langle \theta^{12}[\delta_i], \theta^4[\delta_j] \sum_{\delta'} \theta^8[\delta'], \partial_{\delta_k} I_4 \rangle$	15	$\mathrm{st}_5 \oplus \mathrm{repa}_5$
$\langle \theta^4[\delta_i] \theta^4[\delta_j] \theta^4[\delta_k] \rangle_{\delta_i, \delta_j, \delta_k \text{ pari}}$	35	$S^3 V_\theta$
$\langle \theta^4[\delta_i] \theta^4[\delta_j] \theta^4[\delta_k] \rangle_{\delta_i + \delta_j + \delta_k \text{ pari}}$	35	$S^3 V_\theta$
$\langle \theta^4[\delta_i] \theta^4[\delta_j] \theta^4[\delta_k] \rangle_{\delta_i + \delta_j + \delta_k \text{ dispari}}$	20	$\mathrm{st}_5 \oplus \mathrm{repa}_5 \oplus \mathrm{sw}_{10}$

$\mathbb{H}_2$. This function is a multiple of $\theta[{0 \atop 0}]^4(\tau_1)$. The restrictions of the F_i are also multiples of $\theta[{0 \atop 0}]^4(\tau_1)$, but the restriction of $\sum \theta[\delta]^{16}$ is not. Hence $\Xi_8[0^{(2)}]$ should be linear combination of the three F_i only. We try to determine $a_i \in \mathbb{C}$ such that $\sum_i a_i F_i$ factors in this way for such period matrices. Note that

$$\theta[{ab \atop cd}](\tau_{1,1}) = \theta[{a \atop c}](\tau_1)\theta[{b \atop d}](\tau_1'),$$

where $\tau_{1,1} = \mathrm{diag}(\tau_1, \tau_1')$ and $\tau_1, \tau_1' \in \mathbb{H}_1$. In particular, $\theta[{ab \atop cd}](\tau_{1,1}) = 0$ if $ac = 1$. As $\theta[{00 \atop 00}](\tau_{1,1})$ produces $\theta[{0 \atop 0}]^4(\tau_1)\theta[{0 \atop 0}]^4(\tau_1')$, it remains to find a_i such that

$$\left(\theta[{0 \atop 0}]^4 \eta^{12}\right)(\tau_1) \left(\theta[{0 \atop 0}]^4 \eta^{12}\right)(\tau_1') = (a_1 F_1 + a_2 F_2 + a_3 F_3)(\tau_{1,1}).$$

Using the results from section 5.1., the restrictions of the F_i are

$$\begin{aligned}\theta[^{00}_{00}]^{16}(\tau_{1,1}) &= \theta[^0_0]^{16}(\tau_1)\theta[^0_0]^{16}(\tau_1') \\ &= \theta[^0_0]^4(\tau_1)(\tfrac{1}{3}f_{21}+\eta^{12})(\tau_1) \\ &\quad \cdot \theta[^0_0]^4(\tau_1')(\tfrac{1}{3}f_{21}+\eta^{12})(\tau_1'), \\ \left(\theta[^{00}_{00}]^4\sum_\delta \theta[\delta]^{12})(\tau_{1,1}\right) &= \theta[^0_0]^4(\tau_1)(\theta[^0_0]^{12}+\theta[^0_1]^{12}+\theta[^1_0]^{12})(\tau_1) \\ &\quad \cdot \theta[^0_0]^4(\tau_1')(\theta[^0_0]^{12}+\theta[^0_1]^{12}+\theta[^1_0]^{12})(\tau_1') \\ &= \theta[^0_0]^4(\tau_1)(\tfrac{2}{3}f_{21}-\eta^{12})(\tau_1) \\ &\quad \cdot \theta[^0_0]^4(\tau_1')(\tfrac{2}{3}f_{21}-\eta^{12})(\tau_1'), \\ \left(\theta[^{00}_{00}]^8\sum_\delta \theta[\delta]^8\right)(\tau_{1,1}) &= \left(\theta[^0_0]^8(\theta[^0_0]^8+\theta[^0_1]^8+\theta[^1_0]^8\right)(\tau_1) \\ &\quad \cdot \left(\theta[^0_0]^8(\theta[^0_0]^8+\theta[^0_1]^8+\theta[^1_0]^8)\right)(\tau_1') \\ &= \theta[^0_0]^4(\tau_1)\tfrac{2}{3}f_{21}(\tau_1)\theta[^0_0]^4(\tau_1')\tfrac{2}{3}f_{21}(\tau_1').\end{aligned} \tag{90}$$

Next we require for the term $f_{21}(\tau_1)$ to disappear from the linear combination $(\sum a_iF_i)(\tau_{1,1})$, so that we must have

$$\left(a_1(\tfrac{1}{3}f_{21}+\eta^{12})+2a_2(\tfrac{2}{3}f_{21}-\eta^{12})+2a_3\tfrac{2}{3}f_{21}\right)(\tau_1') = 0$$

for all $\tau_1' \in \mathbb{H}_1$. This gives two linear equations for the a_i which have a unique solution, up to scalar multiple:

$$a_1+4a_2+4a_3=0, \qquad a_1-2a_2=0, \qquad \text{hence} \quad (a_1,a_2,a_3)=\lambda(-4,-2,3).$$

A computation shows that:

$$(-4F_1-2F_2+3F_3)(\tau_{1,1}) = 6\theta[^0_0]^4(\tau_1)\eta^{12}(\tau_1)\theta[^0_0]^4(\tau_1')\eta^{12}(\tau_1').$$

Thus we conclude that

$$\Xi_8[^{00}_{00}] := \theta[^{00}_{00}]^4(-4\theta[^{00}_{00}]^{12}-2\sum_\delta\theta[\delta]^{12}+3\theta[^{00}_{00}]^4\sum_\delta\theta[\delta]^8)/6$$

satisfies the constraints. Because we use a basis for the O^+-invariants and the equations for the a_i's have an unique solution we conclude that the the $\Xi_8[^{00}_{00}]$ is the unique modular form on $\Gamma_2(1,2)$ satisfying the constraints.

As $\theta[^{00}_{00}]^4\Xi_6[^{00}_{00}]$ satisfies the same constraints (with $\Xi_6[^{00}_{00}]$ the modular form determined by D'Hoker and Phong in [DP1], [DP4]) we obtain from uniqueness that

$$\Xi_6[^{00}_{00}] = \left(-4\theta[^{00}_{00}]^{12}-2\sum_\delta\theta[\delta]^{12}+3\theta[^{00}_{00}]^4\sum_\delta\theta[\delta]^8\right)/6.$$

Another formula for this function is:

$$\Xi_6\left[\begin{smallmatrix}00\\00\end{smallmatrix}\right] = -(\theta\left[\begin{smallmatrix}00\\11\end{smallmatrix}\right]\theta\left[\begin{smallmatrix}01\\00\end{smallmatrix}\right]\theta\left[\begin{smallmatrix}10\\01\end{smallmatrix}\right])^4 - (\theta\left[\begin{smallmatrix}00\\01\end{smallmatrix}\right]\theta\left[\begin{smallmatrix}01\\10\end{smallmatrix}\right]\theta\left[\begin{smallmatrix}11\\00\end{smallmatrix}\right])^4 - (\theta\left[\begin{smallmatrix}00\\10\end{smallmatrix}\right]\theta\left[\begin{smallmatrix}10\\00\end{smallmatrix}\right]\theta\left[\begin{smallmatrix}11\\11\end{smallmatrix}\right])^4,$$

which is the one found by D'Hoker and Phong in [DP4]. To check the equality between the two expressions for $\Xi_6\left[\begin{smallmatrix}00\\00\end{smallmatrix}\right]$ one can use the classical theta formula.

We observe that the five dimensional space generated by the $\Xi_6[\delta]$ is exactly the same as the one generated by the derivative of the Igusa quartic w.r.t. to the five p_i (cf. [CD], Theorem 1, for details).

5.3. The case $g = 3$

In this case the decomposition of $\mathrm{Sp}(6)$-representation on $M_{2k}(\Gamma_g(2))$ in irreducible representations is longer than in genus two. So, for sake of clarity, we first construct a modular form Ξ_8 satisfying the three constraints, then we will treat the algebraic proprieties of the representation on the space of modular forms and finally we will prove the uniqueness of Ξ_8.

5.3.1. Modular forms

In case $g = 3$, the 8 $\Theta[\sigma]$'s define a holomorphic map

$$\mathbb{H}_3 \longrightarrow \mathbb{P}^7, \qquad \tau \longmapsto (\Theta[000](\tau) : \ldots : \Theta[111](\tau)). \tag{91}$$

The closure of the image of this map is a 6-dimensional variety which is defined by a homogeneous polynomial F_{16} in eight variables of degree 16, as anticipated in section 3.6.3.. In particular the holomorphic function $\tau \to F_{16}(\cdots, \Theta[\delta], \cdots)$ is identically zero on $\mathbb{H}_3$. It is interesting to review the details of the construction of F_{16} (see [vGvdG]), as we will use an analogous strategy to obtain a certain function, $G[\Delta]$, we will use to define the superstring measures. For all $\tau \in \mathbb{H}_3$, the following relation holds:

$$r_1 - r_2 = r_3, \qquad \text{with} \quad r_1 = \prod_{a,b\in\mathbb{Z}_2} \theta\left[\begin{smallmatrix}000\\0ab\end{smallmatrix}\right](\tau), \tag{92}$$

$$r_2 = \prod_{a,b\in\mathbb{Z}_2} \theta\left[\begin{smallmatrix}000\\1ab\end{smallmatrix}\right](\tau), \qquad r_3 = \prod_{a,b\in\mathbb{Z}_2} \theta\left[\begin{smallmatrix}100\\0ab\end{smallmatrix}\right](\tau).$$

From these we deduce that $2r_1r_2 = r_1^2 + r_2^2 - r_3^2$, and thus

$$r_1^4 + r_2^4 + r_3^4 - 2(r_1^2r_2^2 + r_1^2r_3^2 + r_2^2r_3^2) \tag{93}$$

is zero, as function of τ, on $\mathbb{H}_3$. Let F_{16} be the homogeneous polynomial, of degree 16 in the $\Theta[\sigma]$'s, obtained (using the classical theta formula (54)) from this polynomial (of degree 8) in the $\theta[\Delta]^2$. In [vGvdG] it has been shown that F_{16} is not zero as a polynomial in the eight $\Theta[\sigma]$. Thus the polynomial F_{16} defines the image of $\mathbb{H}_3 \to \mathbb{P}^7$. The same polynomial can be written using the classical theta functions. A computer computation, using once again the classical formula, shows that F_{16} coincides, up to a scalar multiple, with the degree 16 polynomial in the $\Theta[\sigma]$ obtained from

$$8\sum_{\Delta}\theta[\Delta]^{16} - \Big(\sum_{\Delta}\theta[\Delta]^{8}\Big)^{2} \tag{94}$$

by the classical theta formulas.
The polynomial F_{16} provides the only relation of degree 16 between the theta constants and the quotient ring is normal [R1, R2] so we get

$$M_{2k}(\Gamma_3(2)) = M^{\theta}_{2k}(\Gamma_3(2)) = (\mathbb{C}[\ldots,\Theta[\sigma],\ldots]_{4k})^{H_3} \tag{95}$$

where

$$\begin{aligned}&(\mathbb{C}[\ldots,\Theta[\sigma],\ldots]_{4k})^{H_3}\\ &= \begin{cases}(\mathbb{C}[\ldots,X_\sigma,\ldots]_{4k})^{H_3} & k\le 3,\\ (\mathbb{C}[\ldots,X_\sigma,\ldots]_{4k})^{H_3}/F_{16}(\mathbb{C}[\ldots,X_\sigma,\ldots]_{4k-16})^{H_3}, & k\ge 4.\end{cases}\end{aligned} \tag{96}$$

5.3.2. The functions $F_i^{(3)}$

As in the genus two case, we want to find a modular form $\Xi_8\left[\begin{smallmatrix}000\\000\end{smallmatrix}\right]$ of weight 8 on $\Gamma_3(2)$ which restricts to the 'diagonal' $\Delta_{1,2}$ as:

$$\begin{aligned}\Xi_8\left[\begin{smallmatrix}000\\000\end{smallmatrix}\right](\tau_{1,2}) &= \Xi_8\left[\begin{smallmatrix}0\\0\end{smallmatrix}\right](\tau_1)\Xi_8\left[\begin{smallmatrix}00\\00\end{smallmatrix}\right](\tau_2)\\ &= \Big(\theta\left[\begin{smallmatrix}0\\0\end{smallmatrix}\right]^4\eta^{12}\Big)(\tau_1)\Big(\theta\left[\begin{smallmatrix}00\\00\end{smallmatrix}\right]^4\Xi_6\left[\begin{smallmatrix}00\\00\end{smallmatrix}\right]\Big)(\tau_2)\end{aligned} \tag{97}$$

where $\tau_{1,2}\in\mathbb{H}_3$ is the block diagonal matrix with entries $\tau_1\in\mathbb{H}_1$ and $\tau_2\in\mathbb{H}_2$. Obvious generalizations of the functions $F_i^{(2)}$ which we considered in section 5.2.2. are

$$\begin{aligned}F_1^{(3)} &:= \theta\left[\begin{smallmatrix}000\\000\end{smallmatrix}\right]^{16},\\ F_2^{(3)} &:= \theta\left[\begin{smallmatrix}000\\000\end{smallmatrix}\right]^{4}\sum_{\Delta}\theta[\Delta]^{12},\\ F_3^{(3)} &:= \theta\left[\begin{smallmatrix}000\\000\end{smallmatrix}\right]^{8}\sum_{\Delta}\theta[\Delta]^{8},\end{aligned} \tag{98}$$

where the sum is over the 36 even characteristics Δ in genus three. The functions F_i are modular forms of weight 8 on $\Gamma_3(1,2)$, see [CDG1]. We also have the $\mathrm{Sp}(6)$-invariant $\sum_\Delta\theta[\Delta]^{16}$. However, there is no linear combination of these functions which has the desired restriction. Therefore we need another modular form $G\left[\begin{smallmatrix}000\\000\end{smallmatrix}\right]$ of weight 8 on $\Gamma_3(1,2)$. To this end we need the notion of isotropic and Lagrangian subspaces.

5.3.3. Isotropic subspaces

In this section we introduce isotropic subspaces of a space $V \cong \mathbb{Z}_2^{2g}$ for arbitrary g in a quite general approach, as we will use the same notions to tackle the genus four case.
A subspace $W \subset V$ is isotropic if $E(w,w') = 0$ for all $w, w' \in W$. Given a basis $e_1, \dots, e_k$ of W it is not hard to see that one can extend it to a symplectic basis $e_1, \dots e_{2g}$ of V (so $E(e_i, e_j) = 0$ unless $|i-j| = g$ and then $E(e_i, e_j) = 1$). In particular, the group $Sp(2g, \mathbb{Z})$ acts transitively on the isotropic subspaces of V of a given dimension. The number of k-dimensional isotropic subspaces of $V \cong \mathbb{Z}_2^{2g}$ is given by

$$\begin{aligned} N_{iso}(g,k) &= \frac{(2^{2g}-1)(2^{2g-1}-2)(2^{2g-4}-4)\dots(2^{2g-(k-1)}-2^{k-1})}{\sharp GL(k,\mathbb{Z}_2)} \qquad (99)\\ &= \frac{(2^{2g}-1)(2^{2g-1}-2)(2^{2g-2}-4)\dots(2^{2g-(k-1)}-2^{k-1})}{(2^k-1)(2^k-2)\dots(2^k-2^{k-1})} \\ &= \frac{(2^{2g}-1)(2^{2g-2}-1)(2^{2g-4}-1)(2^{2g-6}-1)\dots(2^{2(g-k)+2}-1)}{(2^k-1)(2^{k-1}-1)\dots(2-1)}, \end{aligned}$$

where in the numerator we count the ordered k-tuples of independent elements $v_1, \dots, v_k \in V$ with $E(v_i, v_j) = 0$ for all i, j: for v_1 we can take any element in $V - \{0\}$, for v_2 we can take any element in $\langle v_1\rangle^\perp \cong \mathbb{Z}_2^{2g-1}$ except $0, v_1$, so $v_2 \in \langle v_1\rangle^\perp - \langle v_1\rangle$, next $v_3 \in \langle v_1, v_2\rangle^\perp - \langle v_1, v_2\rangle$ and so on.

Let $W_1, \dots, W_N$ be the k-dimensional isotropic subspaces contained in an even quadric $Q \subset V$ defined by $q = 0$, where q is a quadratic form on V. Let $\sigma \in Sp(2g, \mathbb{Z})$ be a symplectic transformation. Then $\sigma(W_1), \dots, \sigma(W_N)$ are the k-dimensional isotropic subspaces in the even quadric $\sigma(Q) \subset V$ defined by $\sigma \cdot q = 0$, with $(\sigma \cdot q)(\sigma v) = q(v)$. In particular, all even quadrics in V contains the same number of isotropic subspaces of a given dimension. An even quadric contains a maximal isotropic subspace L: this is an isotropic subspace not contained in any higher dimensional nontrivial isotropic subspace. By trivial we mean the space containing the 0 only or the whole V. It follows that maximal isotropic subspaces are half dimensional (t.i. g dimensional) subspaces. For example, $L_0 = \{\binom{v_1 \dots v_g}{0 \dots 0} : v_i \in \mathbb{Z}_2\}$ is contained in the even quadric Q corresponding to the characteristic $\left[\begin{smallmatrix}0\dots0\\0\dots0\end{smallmatrix}\right]$. Instead, odd quadrics do not contain maximal isotropic subspaces: if $L \subset Q$ were such a subspace, then, by transitivity, $\sigma(L) = L_0$ for a suitable $\sigma \in Sp(2g, \mathbb{Z})$. If $\sigma(Q)$ corresponds to the characteristic $\left[\begin{smallmatrix}a\\b\end{smallmatrix}\right]$ then $L_0 \subset \sigma(Q)$ implies $a_1 = \dots = a_g = 0$, hence the characteristic must be even. A maximally isotropic subspace, L, is called a *Lagrangian*. For example, if $g = 3$, a subspace L of $V \cong \mathbb{Z}_2^6$ is Lagrangian if $L \subset V$, $E(v,w) = 0$ for all $v, w \in L$ and $\dim L = 3$. The eight elements $\binom{abc}{000} \in V$ with $a, b, c \in \mathbb{Z}_2$ form a Lagrangian subspace L_0 in V. Instead, the higher dimensional isotropic subspaces contained in an odd quadric are $(g-1)$-dimensional. For example, $W_0 = \{\binom{v_1 \dots v_{g-1} 0}{0 \dots 0 \quad 0} : v_i \in \mathbb{Z}_2\}$ is contained in the odd quadric with characteristic $\left[\begin{smallmatrix}0\dots01\\0\dots01\end{smallmatrix}\right]$.

It is easy to count the number of even quadrics which contain a fixed k-dimensional isotropic subspace: we may assume that the subspace has basis $e_1, \dots, e_k$ so that the characteristic of an even quadric containing it is

$$\begin{bmatrix} 0 & \dots & 0 & a_{k+1} & \dots & a_g \\ b_1 & \dots & b_k & b_{k+1} & \dots & b_g \end{bmatrix},$$

with $\sum_{i=k+1}^{g} a_{k+i}b_{k+i} = 0$, and the number of such even quadrics is

$$N_Q(k) = 2^k \cdot 2^{g-k-1}(2^{g-k}+1). \tag{100}$$

Viceversa, to find the number of k-dimensional isotropic subspaces in an even quadric, we can count the pairs (W, Q) of isotropic subspaces W contained in arbitrary even quadrics Q in two ways: first as the product of the number of W with the number of even Q containing a fixed W and second as the product of the number of even quadrics with the number of k-dimensional isotropic subspaces in an even quadric. For example, let us consider maximal isotropic subspaces. Then $g = k$ and the total number of copies (W, Q) is in this case

$$N = N_{iso}(g, g)N_Q(g) = [(2^g+1)(2^{g-1}+1)\dots(2+1)]2^g.$$

On the other side we know that maximally isotropic spaces are contained in even quadrics only, which are in one-to-one correspondence with the set of even characteristics, which are $N_e = 2^{g-1}(2^g+1)$. Thus, we conclude that the number of maximally isotropic subspaces contained in a given (even) quadric Q is

$$N_{iso}^{Q}(g) = N_{iso}(g, g)N_Q(g)/N_e = 2[(2^{g-1}+1)\dots(2+1)]. \tag{101}$$

For example, the number of pairs (W, Q) of a maximally isotropic subspace in an even quadric in $\mathbb{Z}_2^6$ is $135 \cdot 2^3$, and thus the number of such subspaces in a fixed Q is $135 \cdot 2^3/36 = 30$. The general idea to count the number of k-dimensional isotropic subspaces in an even quadric is graphically pictured in Figure 1: it is clear that in the first way we count the couples by columns, and in the second way by rows.

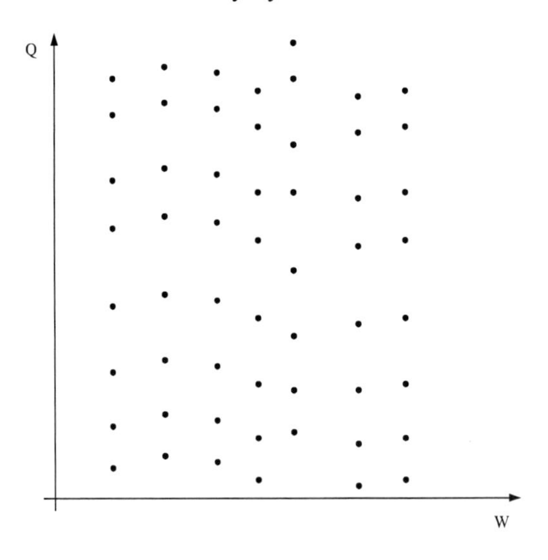

Figure 1. Counting of the couple $(Q,\ W)$

For small g we list some of these dimensions in the table on the left below, in the table on the right we list the number of k-dimensional isotropic subspaces contained in an even quadric.

$g \setminus$ dimension	1	2	3	4
1	3			
2	15	15		
3	63	315	135	
4	255	5355	11475	2295

$g \setminus$ dimension	1	2	3	4
1	2			
2	9	6		
3	35	105	30	
4	135	1575	2025	270

5.3.4. The modular forms $G[\Delta]$

For each even characteristic Δ in $g = 3$ we define a modular form $G[\Delta]$ of weight 8 on $\Gamma_3(2)$. First, let us recall some facts and notations about characteristics and quadrics introduced in section 3.2., specializing to the case $g = 3$. An even characteristic Δ corresponds to a quadratic form

$$q_\Delta : V = \mathbb{Z}_2^6 \longrightarrow \mathbb{Z}_2$$

which satisfies $q_\Delta(v+w) = q_\Delta(v)+q_\Delta(w)+E(v,w)$, where $E(v,w) := \sum_{i=1}^{3}(v_i w_{3+i} + v_{3+i}w_i)$. If $\Delta = \left[\begin{smallmatrix} abc \\ def \end{smallmatrix}\right]$ then

$$q_\Delta(v) = v_1v_4 + v_2v_5 + v_3v_6 + av_1 + bv_2 + cv_3 + dv_4 + ev_5 + fv_6,$$

with $v = (v_1, \ldots, v_6) \in V$. We will also write $v = \left(\begin{smallmatrix} v_1v_2v_3 \\ v_4v_5v_6 \end{smallmatrix}\right)$. Let $Q_\Delta = \{v \in V : q_\Delta(v) = 0\}$ be the corresponding quadric in V.

Let L be a Lagrangian subspace of V. For such a subspace L we define a modular form on a subgroup of $Sp(6,\mathbb{Z})$:

$$P_L := \prod_{Q \supset L} \theta[\Delta_Q]^2$$

here the product is over all even quadrics which contain L (there are eight such quadrics for each L, as explained in the previous Section) and Δ_Q is the even characteristic corresponding to Q. In case $L = L_0$ with

$$L_0 := \{(v_1, \ldots, v_6) \in V : v_4 = v_5 = v_6 = 0\},$$

we have

$$P_{L_0} = (2r_1r_2)^2 = \prod_{a,b,c \in \mathbb{Z}_2} \theta\left[\begin{smallmatrix} 000 \\ abc \end{smallmatrix}\right]^2,$$

with r_1, r_2 as in section 5.3.1.. The action of $Sp(6,\mathbb{Z})$ on $V = \mathbb{Z}^6/2\mathbb{Z}^6$ permutes the Lagrangian subspaces L, and the subgroup $\Gamma_3(2)$ acts trivially on V. Similarly, the P_L are permuted by the action of $Sp(6,\mathbb{Z})$, see [CDG1], and as $\Gamma_3(2)$ fixes all L's, the P_L are modular forms on $\Gamma_3(2)$ of weight 8.

For an even characteristic Δ, the quadric Q_Δ contains 30 Lagrangian subspaces. The sum of the 30 P_L's, with L a Lagrangian subspace of Q_Δ, is a modular form $G[\Delta]$ of weight 8 on $\Gamma_3(2)$:

$$G[\Delta] := \sum_{L \subset Q_\Delta} P_L = \sum_{L \subset Q_\Delta} \prod_{Q' \supset L} \theta[\Delta_{Q'}]^2.$$

Note that $\theta[\Delta]^2$ is one of the factors in each of the 30 products. As the P_I are permuted by the action of $Sp(6,\mathbb{Z})$, also the $G[\Delta]$ are permuted:

$$G[M\cdot\Delta](M\cdot\tau) = \det(C\tau+D)^8 G[\Delta](\tau).$$

Since $\Gamma_3(1,2)$ fixes the characteristic $\left[\begin{smallmatrix}000\\000\end{smallmatrix}\right]$, the function $G\left[\begin{smallmatrix}000\\000\end{smallmatrix}\right]$ is a modular form on $\Gamma_3(1,2)$.

5.3.5. The restriction

Now, we can tackle the problem of finding a linear combination of the functions F_i, $i = 1,2,3$ and $G\left[\begin{smallmatrix}000\\000\end{smallmatrix}\right]$, in order to satisfy the third constraint:

$$\left(\theta\left[\begin{smallmatrix}0\\0\end{smallmatrix}\right]^4\eta^{12}\right)(\tau_1)\left(\theta\left[\begin{smallmatrix}00\\00\end{smallmatrix}\right]^4\Xi_6\left[\begin{smallmatrix}00\\00\end{smallmatrix}\right]\right)(\tau_2) = \left(b_1F_1+b_2F_2+b_3F_3+b_4G\left[\begin{smallmatrix}000\\000\end{smallmatrix}\right]\right)(\tau_{1,2}).$$

It is easy to see that the theta constants satisfy

$$\theta\left[\begin{smallmatrix}abc\\def\end{smallmatrix}\right](\tau_{1,2}) = \theta\left[\begin{smallmatrix}a\\d\end{smallmatrix}\right](\tau_1)\theta\left[\begin{smallmatrix}bc\\ef\end{smallmatrix}\right](\tau_2),$$

so that, in particular, $\theta\left[\begin{smallmatrix}abc\\def\end{smallmatrix}\right]\mapsto 0$ when $ad=1$. As $\left[\begin{smallmatrix}abc\\def\end{smallmatrix}\right]$ must be even, this happen for $\left[\begin{smallmatrix}bc\\ef\end{smallmatrix}\right]$ odd, thus 6 of the 36 even theta constants map to zero. The other $30 = 3\cdot 10$ are uniquely decomposed in the product of two even theta constants for $g=1$ (3) and $g=2$ (10) respectively. Using the results from 5.1., the functions $F_i(\tau_{1,2})$ are easily made explicit, and the function $G\left[\begin{smallmatrix}000\\000\end{smallmatrix}\right](\tau_{1,2})$ has been determined in [CDG1]. The restrictions to $\Delta_{1,2}\cong \mathbb{H}_1\times\mathbb{H}_2$ are:

$$\begin{aligned}
(\theta\left[\begin{smallmatrix}000\\000\end{smallmatrix}\right]^{16})(\tau_{1,2}) &= \theta\left[\begin{smallmatrix}0\\0\end{smallmatrix}\right]^{16}(\tau_1)\theta\left[\begin{smallmatrix}00\\00\end{smallmatrix}\right]^{16}(\tau_2)\\
&= \left(\theta\left[\begin{smallmatrix}0\\0\end{smallmatrix}\right]^4(\tfrac{1}{3}f_{21}+\eta^{12})\right)(\tau_1)\theta\left[\begin{smallmatrix}00\\00\end{smallmatrix}\right]^{16}(\tau_2),\\
\theta\left[\begin{smallmatrix}000\\000\end{smallmatrix}\right]^4(\sum_\Delta\theta[\Delta]^{12})(\tau_{1,2}) &= \left(\theta\left[\begin{smallmatrix}0\\0\end{smallmatrix}\right]^4(\theta\left[\begin{smallmatrix}0\\0\end{smallmatrix}\right]^{12}+\theta\left[\begin{smallmatrix}0\\1\end{smallmatrix}\right]^{12}+\theta\left[\begin{smallmatrix}1\\0\end{smallmatrix}\right]^{12})\right)(\tau_1)\\
&\qquad\cdot\left(\theta\left[\begin{smallmatrix}00\\00\end{smallmatrix}\right]^4(\sum_\delta\theta[\delta]^{12})\right)(\tau_2)\\
&= \left(\theta\left[\begin{smallmatrix}0\\0\end{smallmatrix}\right]^4(\tfrac{2}{3}f_{21}-\eta^{12})\right)(\tau_1)\left(\theta\left[\begin{smallmatrix}00\\00\end{smallmatrix}\right]^4\sum_\delta\theta[\delta]^{12}\right)(\tau_2),\\
(\theta\left[\begin{smallmatrix}000\\000\end{smallmatrix}\right]^8\sum_\Delta\theta[\Delta]^8)(\tau_{1,2}) &= \left(\theta\left[\begin{smallmatrix}0\\0\end{smallmatrix}\right]^8(\theta\left[\begin{smallmatrix}0\\0\end{smallmatrix}\right]^8+\theta\left[\begin{smallmatrix}0\\1\end{smallmatrix}\right]^8+\theta\left[\begin{smallmatrix}1\\0\end{smallmatrix}\right]^8)\right)(\tau_1)\\
&\qquad\cdot\left(\theta\left[\begin{smallmatrix}00\\00\end{smallmatrix}\right]^8(\sum_\delta\theta[\delta]^8)\right)(\tau_2)\\
&= \left(\theta\left[\begin{smallmatrix}0\\0\end{smallmatrix}\right]^4\tfrac{2}{3}f_{21}\right)(\tau_1)\left(\theta\left[\begin{smallmatrix}00\\00\end{smallmatrix}\right]^8(\sum_\delta\theta[\delta]^8)\right)(\tau_2),\\
G\left[\begin{smallmatrix}000\\000\end{smallmatrix}\right](\tau_{1,2}) &= \left(\theta\left[\begin{smallmatrix}0\\0\end{smallmatrix}\right]^4(\tfrac{1}{3}f_{21}-\eta^{12})\right)(\tau_1)\\
&\qquad\cdot\left(\theta\left[\begin{smallmatrix}00\\00\end{smallmatrix}\right]^4(\tfrac{1}{3}\theta\left[\begin{smallmatrix}00\\00\end{smallmatrix}\right]^{12}+\tfrac{2}{3}\sum_\delta\theta[\delta]^{12}-\tfrac{1}{2}\theta\left[\begin{smallmatrix}00\\00\end{smallmatrix}\right]^4\sum_\delta\theta[\delta]^8)\right)(\tau_2).
\end{aligned}$$

In putting these expression in $(b_1F_1+b_2F_2+b_3F_3)+b_4G[^{000}_{000}]$, we note that the common function $\theta[^{000}_{000}]^4$ in front of the F_i gives the function $\theta[^0_0]^4(\tau_1)\theta[^{00}_{00}]^4(\tau_2)$. In particular, the restriction has then a factor $\theta[^0_0]^4\theta[^{00}_{00}]^4$. In order for this restriction to be a multiple of $\theta[^0_0]^4\eta^{12}$, the terms containing f_{21} must disappear. This leads to the equation

$$b_1\theta[^{00}_{00}]^{12}+2b_2\sum_\delta\theta[\delta]^{12}+2b_3\theta[^{00}_{00}]^4(\sum_\delta\theta[\delta]^8)+b_4(\tfrac{1}{3}\theta[^{00}_{00}]^{12}+\tfrac{2}{3}\sum_\delta\theta[\delta]^{12}-\tfrac{1}{2}\theta[^{00}_{00}]^4\sum_\delta\theta[\delta]^8)=0.$$

It has a unique solution (up to scalar multiples):

$$(b_1,b_2,b_3,b_4)\;=\;\mu(4,4,-3,-12)\qquad(\mu\in\mathbb{C}).$$

Setting $\mu=1$ and using the formula for $\Xi_6[^{00}_{00}]$ given in section 5.2.2. we get

$$\begin{aligned}\Big(4F_1+4F_2-3F_3-12G[^{000}_{000}]\Big)(\tau_{1,2}) &=\\ \Big(\theta[^0_0]^4\eta^{12}\Big)(\tau_1)\Big(\theta[^{00}_{00}]^4(8\theta[^{00}_{00}]^{12}+4\textstyle\sum_\delta\theta[\delta]^{12}-6\theta[^{00}_{00}]^4\sum_\delta\theta[\delta]^8)\Big)(\tau_2) &=\\ 12\Big(\theta[^0_0]^4\eta^{12}\Big)(\tau_1)\Big(\theta[^{00}_{00}]^4\Xi_6[^{00}_{00}]\Big)(\tau_2).\end{aligned}$$

Hence the modular form $\Xi_8[^{000}_{000}]$, of weight 8 on $\Gamma_3(1,2)$ defined by

$$\Xi_8[^{000}_{000}]\;:=\;(4F_1+4F_2-3F_3-12G[^{000}_{000}])/12$$

satisfies all the constraints except maybe (iii$_0$)(2). We have to check this last constraint separately. Let $M\in\mathrm{Sp}(6,\mathbb{Z})$ be such that $M\cdot[^{000}_{000}]=[^{abc}_{def}]$ with $a=d=1$. As $G[^{000}_{000}](\tau)=\theta^2[^{000}_{000}](\tau)G^b[^{000}_{000}](\tau)$ for a holomorphic function $G^b[^{000}_{000}]$, $G[^{abc}_{def}](M\cdot\tau)$ is the product of $\theta^2[^{abc}_{def}](\tau)$ and a holomorphic function, hence $G[^{abc}_{def}](M\cdot\tau_{1,2})=0$ because $\theta^2[^{abc}_{def}](\tau_{1,2})=\theta^2[^1_1](\tau_1)\theta^2[^{bc}_{ef}](\tau_2)=0$.

We conclude that $\Xi_8[^{000}_{000}]$, defined as above, solves the problem.

Next, using the representation theory of finite group, we will show that, up to a constant, it is the only modular form of weight 8 on $\Gamma_3(1,2)$ which satisfies all constraints. As a by product, this implies that the desired functions $\Xi_6[\Delta]$ proposed by D'Hoker and Phong in [DP6] indeed do not exist because $G[^{000}_{000}]$ is not the product of $\theta[^{000}_{000}]^4$ with a modular form of weight 6. As noticed by Morozov, it has been in part the convincement that the factorization of the term $\theta[^{000}_{000}]^4$ was necessary to grant the vanishing of the cosmological constant to stop for a long time the success in finding higher loop candidates for the superstring measure. Instead, we will see that $\Xi_8[^{000}_{000}]$ implies the vanishing of the cosmological constant.

5.4. Representations of $\mathrm{Sp}(6)$

To prove uniqueness of the form Ξ_8, we need to come back to the study of the representations of $\mathrm{Sp}(6)$ on the space of modular forms.

5.4.1. Weight 2

From section 3.8.4. we know that $M_2((\Gamma_3(2))$, a fifteen dimensional vector space, is an irreducible $Sp(6)$-representation, denoted by ρ_θ. As $Sp(6)$ has a unique irreducible representation of dimension 15, denoted by $\mathbf{15}_a$ in [F1] and in Table 11, it follows that $\rho_\theta\cong\mathbf{15}_a$.

Table 11. Table of characters of $\mathrm{Sp}(6)$. In the bold column there are the characters of the class of the transvections.

Irrep./Class																														
$\mathbf{1}_a$	1	1	1	1	1	1	1	1	1	1	1	1	1	1	1	**1**	1	1	1	1	1	1	1	1	1	1	1	1	1	1
$\mathbf{7}_a$	7	3	-1	3	1	4	2	0	1	-1	-2	2	0	1	2	**-5**	-1	-3	1	-1	-2	1	-1	-2	0	-1	0	-1	1	0
$\mathbf{27}_a$	27	7	3	3	1	9	3	1	0	0	0	0	0	0	2	**15**	3	5	1	1	3	0	0	1	0	-1	-1	-1	-1	-1
$\mathbf{21}_a$	21	1	-3	5	-1	6	0	-2	0	0	3	3	-1	0	1	**9**	-3	3	-1	-1	0	0	0	2	-1	1	0	1	1	0
$\mathbf{35}_a$	35	-5	3	7	-1	5	-3	1	2	0	-1	3	1	-1	0	**-5**	3	-1	-1	1	1	-2	0	-1	0	0	-1	-1	1	0
$\mathbf{105}_a$	105	5	1	5	-1	15	1	-1	-3	1	-3	1	-1	0	0	**-35**	1	-5	-1	1	1	1	1	-1	0	0	1	1	-1	0
$\mathbf{189}_a$	189	-11	-3	9	1	9	-3	1	0	0	0	0	0	0	-1	**21**	-3	1	1	-1	-3	0	0	1	1	-1	1	1	-1	0
$\mathbf{21}_b$	21	5	5	1	1	6	2	2	0	2	3	-1	1	0	1	**-11**	-3	-3	-3	-1	-2	-2	0	0	-1	1	0	1	-1	0
$\mathbf{35}_b$	35	7	11	-1	1	5	-1	1	2	2	-1	-1	-1	-1	0	**15**	3	1	5	-1	3	0	0	-1	0	0	1	3	1	0
$\mathbf{189}_b$	189	13	-3	-3	1	9	-3	1	0	0	0	0	0	0	-1	**-51**	-3	1	1	1	-3	0	0	1	-1	-1	1	-3	1	0
$\mathbf{189}_c$	189	1	21	-3	-1	9	3	1	0	0	0	0	0	0	-1	**-39**	-3	-1	-5	-1	3	0	0	1	1	-1	-1	1	1	0
$\mathbf{15}_a$	15	3	7	-1	1	0	-2	0	3	1	-3	1	-1	0	0	**-5**	-1	1	-3	1	-2	1	-1	0	0	0	-2	3	-1	1
$\mathbf{105}_b$	105	9	-7	-3	1	0	-4	0	3	-1	6	2	0	0	0	**25**	1	-3	-3	-1	4	1	1	0	0	0	0	-3	-1	0
$\mathbf{105}_c$	105	-3	17	-3	-1	0	2	0	3	-1	6	2	0	0	0	**5**	-7	-1	3	1	2	-1	-1	0	0	0	2	1	-1	0
$\mathbf{315}_a$	315	3	-21	-5	-1	0	0	0	0	0	-9	3	1	0	0	**-45**	3	3	3	-1	0	0	0	0	0	0	0	3	-1	0
$\mathbf{405}_a$	405	-3	-27	-3	1	0	0	0	0	0	0	0	0	0	0	**45**	-3	-3	-3	1	0	0	0	0	0	0	0	5	1	-1
$\mathbf{168}_a$	168	8	8	0	0	6	2	2	-3	-1	6	2	0	0	-2	**40**	8	0	0	0	-2	1	-1	0	0	1	0	0	0	0
$\mathbf{56}_a$	56	8	-8	0	0	11	1	-1	2	-2	2	-2	0	-1	1	**-24**	0	-4	4	0	-3	0	0	1	1	1	-1	0	0	0
$\mathbf{120}_a$	120	8	-8	0	0	15	1	-1	0	-2	-6	-2	0	0	0	**40**	0	4	-4	0	1	-2	0	-1	0	0	1	0	0	1
$\mathbf{210}_a$	210	2	2	-2	-2	15	-1	-1	0	2	3	-1	1	0	0	**50**	-6	2	2	0	-1	2	0	-1	0	0	-1	-2	0	0
$\mathbf{280}_a$	280	-8	-8	0	0	10	-2	-2	1	1	10	-2	0	1	0	**-40**	8	0	0	0	2	-1	-1	0	0	0	0	0	0	0
$\mathbf{336}_a$	336	-16	16	0	0	6	-2	2	0	-2	-6	-2	0	0	1	**-16**	0	0	0	0	2	2	0	0	-1	1	0	0	0	0
$\mathbf{216}_a$	216	8	24	0	0	-9	-3	-1	0	0	0	0	0	0	1	**-24**	0	4	-4	0	-3	0	0	-1	1	1	1	0	0	-1
$\mathbf{512}_a$	512	0	0	0	0	-16	0	0	-4	0	8	0	0	-1	2	**0**	0	0	0	0	0	0	0	0	0	-1	0	0	0	1
$\mathbf{378}_a$	378	2	-6	6	2	-9	3	-1	0	0	0	0	0	0	-2	**-30**	-6	2	2	0	3	0	0	-1	0	1	-1	-2	0	0
$\mathbf{84}_a$	84	4	20	4	0	-6	2	-2	3	-1	3	-1	1	0	-1	**4**	4	0	0	0	-2	1	1	0	-1	-1	0	4	0	0
$\mathbf{420}_a$	420	-12	4	-4	0	0	4	0	3	1	-3	1	-1	0	0	**20**	4	0	0	0	-4	-1	1	0	0	0	0	-4	0	0
$\mathbf{280}_b$	280	8	24	0	0	-5	-3	-1	-2	0	-8	0	0	1	0	**40**	0	-4	4	0	1	-2	0	1	0	0	-1	0	0	0
$\mathbf{210}_b$	210	10	-14	6	-2	-15	1	1	3	1	-6	-2	0	0	0	**10**	2	-2	-2	0	1	1	-1	1	0	0	1	-2	0	0
$\mathbf{70}_a$	70	6	-10	2	-2	-5	-1	3	1	-1	7	-1	-1	1	0	**-10**	-2	2	2	0	-1	-1	1	-1	0	0	-1	2	0	0

In the spirit of the genus two case, this space of Heisenberg invariants has basis

$$
\begin{aligned}
P_0 &= \Theta[000]^4 + \Theta[001]^4 + \Theta[010]^4 + \Theta[011]^4 + \Theta[100]^4 + \Theta[101]^4 + \Theta[110]^4 + \Theta[111]^4 \\
P_1 &= 2(\Theta[000]^2\Theta[001]^2 + \Theta[010]^2\Theta[011]^2 + \Theta[100]^2\Theta[101]^2 + \Theta[110]^2\Theta[111]^2) \\
P_2 &= 2(\Theta[000]^2\Theta[010]^2 + \Theta[001]^2\Theta[011]^2 + \Theta[100]^2\Theta[110]^2 + \Theta[101]^2\Theta[111]^2) \\
P_3 &= 2(\Theta[000]^2\Theta[011]^2 + \Theta[001]^2\Theta[010]^2 + \Theta[100]^2\Theta[111]^2 + \Theta[101]^2\Theta[110]^2) \\
P_4 &= 2(\Theta[000]^2\Theta[100]^2 + \Theta[001]^2\Theta[101]^2 + \Theta[010]^2\Theta[110]^2 + \Theta[011]^2\Theta[111]^2) \\
P_5 &= 2(\Theta[000]^2\Theta[101]^2 + \Theta[001]^2\Theta[100]^2 + \Theta[010]^2\Theta[111]^2 + \Theta[011]^2\Theta[110]^2) \\
P_6 &= 2(\Theta[000]^2\Theta[110]^2 + \Theta[001]^2\Theta[111]^2 + \Theta[010]^2\Theta[100]^2 + \Theta[011]^2\Theta[101]^2) \\
P_7 &= 2(\Theta[000]^2\Theta[111]^2 + \Theta[001]^2\Theta[110]^2 + \Theta[010]^2\Theta[101]^2 + \Theta[100]^2\Theta[011]^2) \\
P_8 &= 4(\Theta[000]\Theta[001]\Theta[010]\Theta[011] + \Theta[100]\Theta[101]\Theta[110]\Theta[111]) \\
P_9 &= 4(\Theta[000]\Theta[001]\Theta[100]\Theta[101] + \Theta[010]\Theta[011]\Theta[110]\Theta[111]) \\
P_{10} &= 4(\Theta[000]\Theta[001]\Theta[110]\Theta[111] + \Theta[010]\Theta[011]\Theta[100]\Theta[101]) \\
P_{11} &= 4(\Theta[000]\Theta[010]\Theta[100]\Theta[110] + \Theta[001]\Theta[011]\Theta[101]\Theta[111]) \\
P_{12} &= 4(\Theta[000]\Theta[010]\Theta[101]\Theta[111] + \Theta[001]\Theta[011]\Theta[100]\Theta[110]) \\
P_{13} &= 4(\Theta[000]\Theta[011]\Theta[100]\Theta[111] + \Theta[001]\Theta[010]\Theta[101]\Theta[110]) \\
P_{14} &= 4(\Theta[000]\Theta[011]\Theta[101]\Theta[110] + \Theta[001]\Theta[010]\Theta[100]\Theta[111]).
\end{aligned}
$$

This has been constructed following sec. 3.6.: let us fix four numbers $\sigma_1, \ldots, \sigma_4 \mathbb{Z}_2^3$ which sum up to 0. Then an invariant element is

$$P_{\{\sigma_1,\ldots,\sigma_4\}} = \sum_{x\in\mathbb{Z}^3} \Theta[\sigma_1 + x]\Theta[\sigma_2 + x]\Theta[\sigma_3 + x]\Theta[\sigma_4 + x].$$

Looking at all possible quadruples of numbers σ_i we easily recover the above basis. Using the classical formula one shows that this space is also spanned by the 36 $\theta[\Delta]^4$'s with Δ even.

5.4.2. Weight 4

From section 3.8.4. we know that $\mathrm{Sym}^2(M_2(\Gamma_3(2))) \subset M_4(\Gamma_3(2))$ and as an Sp(6)-representation we have, with an analogous computation as in section 5.2.1.:

$$\mathrm{Sym}^2(M_2(\Gamma_3(2))) := \mathrm{Sym}^2(\mathbf{15}_a) = \mathbf{1} + \mathbf{35}_b + \mathbf{84}_a.$$

The invariant subspace is spanned by $\sum_\Delta \theta[\Delta]^8$ and the subrepresentation $\mathbf{1} + \mathbf{35}_b$ is spanned by the 36 $\theta[\Delta]^8$'s which are permuted by Sp(6).

We recall the relation, introduced in Section 5.3.1., which holds for all $\tau \in \mathbb{H}_3$:

$$r_1 - r_2 = r_3, \qquad \text{with} \quad r_1 = \prod_{a,b\in\mathbb{Z}_2} \theta\begin{bmatrix}000\\0ab\end{bmatrix}(\tau), \quad r_2 = \prod_{a,b\in\mathbb{Z}_2} \theta\begin{bmatrix}000\\1ab\end{bmatrix}(\tau), \quad r_3 = \prod_{a,b\in\mathbb{Z}_2} \theta\begin{bmatrix}100\\0ab\end{bmatrix}(\tau).$$

From this we deduce that $2r_1r_2 = r_1^2 + r_2^2 - r_3^2$. Thus r_1r_2, a product of 8 distinct $\theta[\Delta]$'s, is a linear combination of three products of four theta squares. The sum of the four characteristics in each product is zero, hence

$$r_1r_2 = \prod_{a,b,c\in\mathbb{Z}_2} \theta\begin{bmatrix}000\\abc\end{bmatrix} \in M_4(\Gamma_3(2)).$$

Using a computer and the classical theta formula, we verified that under the action of Sp(6) on r_1r_2 one obtains 135 functions which are a basis of $M_4(\Gamma_3(2))$ and which are permuted (without signs) by Sp(6).

Let $P \subset \mathrm{Sp}(6)$ be the stabilizer of r_1r_2, it consists of the matrices with blocks $A, \ldots, D$ with $C = 0$. There are no non-trivial homomorphisms $P \to GL_1(\mathbb{C}) = \mathbb{C}^*$, because these factor over $SL(3, \mathbb{Z}_2)$ (with P as before, one maps the matrix first to A) and this is a simple group (of order 168). Thus any $g \in P$ acts as the identity on r_1r_2. Acting with the whole group Sp(6) we generate the induced representation of the $\mathbf{1_P}$, by Proposition **??**. We can then identify the representation of Sp(6) on $M_4(\Gamma_3(2))$ with $\mathrm{Ind}_P^{\mathrm{Sp}}(1_P)$, this representation is (cf. [F1], p. 113)

$$M_4(\Gamma_3(2)) \cong \mathrm{Ind}_P^{\mathrm{Sp}}(1_P) \cong \mathbf{1} + \mathbf{35}_b + \mathbf{84}_a + \mathbf{15}_a \cong \mathrm{Sym}^2(M_2(\Gamma_3(2))) + \mathbf{15}_a.$$

In particular, there is a unique complementary 15-dimensional subspace which is Sp(6)-invariant. Necessarily, the representation on this subspace must be $\mathbf{15}_a \cong \rho_\theta$. We can realize an isomorphism of representations, by using some geometry of quadrics, as follows.

Let $L \subset \mathbb{Z}_2^6$ be a Lagrangian subspace: $L \cong \mathbb{Z}_2^3$ and $E(v,w) = 0$ for all $v, w \in L$. Then we know that there are $N_Q(3) = 8$ even quadrics Q such that $L \subset Q$. When $L = L_0 = \{(v, v') \in \mathbb{Z}^3 \times \mathbb{Z}^3 : v' = 0\}$, the 8 even quadrics containing L_0 have the same characteristics as the eight theta constants in $r_1 r_2$ so that we can write

$$r_1 r_2 = \prod_{Q \supset L_0} \theta[\Delta_Q],$$

the product being on the quadrics containing L_0. Acting with the modular group $\mathrm{Sp}(6)$ will generate $N_{iso}(3,3) = 135$ elements of the orbit which are naturally identified, by construction, with the elements in the orbit of $r_1 r_2$, so that each one of them can be written as

$$P_L = \pm \prod_{Q \supset L} \theta[\Delta_Q],$$

for a unique Lagrangian subspace L, where the sign is determined by the condition that it must be $+1$ if $L = L_0$ and that $P_L = \rho(g) P_{L_0}$ for some $g \in \mathrm{Sp}(6)$, with ρ defining a representation.
We can exploit this description for the basis of $M_4(\Gamma_3(2))$ to write down the (unique) O^+-anti-invariant. Recall that O^+ is the stabilizer of the even characteristic $[0]$ and let us call Q_0 the corresponding even quadric. An even (also called split) quadric Q in $\mathbb{Z}_2^6$ contains $N_{iso}^Q(3) = 30$ Lagrangian subspaces. It is easy to see that the intersection of two distinct Lagrangian subspaces in the same quadric cannot have codimension smaller then $g+2$. For $g = 3$ this means that the intersection between two distinct Lagrangian subspaces must have dimension 0 or 1. We say that $L, L' \subset Q$ are in the same ruling if $L \cap L'$ is 1-dimensional. As fixing a point in Q one can easily count 15 Lagrangian subspaces containing it, we see that there are two rulings, each one containing 15 Lagrangian subspaces. Let us define $P[0]$ to be the sum of the 15 P_L's from one ruling minus the sum of the 15 P_L's from the other ruling of Q_0. Then $P[0]$ transforms with the representation ϵ of O^+. Indeed, the action of $\mathrm{Sp}(6)$ preserves intersection so that a given elements $h \in \mathrm{Sp}(6)$ or acts separately on each rule and then as the identity on $P[0]$, or mix the rules acting as -1 on $P[0]$ so that $P[0]$ supports a 1-dimensional non-trivial representation of O^+. But we have said in section 3.8. that ϵ is the unique non-trivial representation.
Thus the subrepresentation of $M_4(\Gamma_3(2))$ generated by $P[0]$ is contained in $\mathrm{Ind}_{O^+}^{\mathrm{Sp}}(\epsilon)$ and thus it must be $\rho_\theta = \mathbf{15}_a$, $\rho_r = \mathbf{21}_b$ or their direct sum. As only $\mathbf{15}_a$ is a component of the representation on $M_4(\Gamma_3(2))$, we conclude that the subrepresentation generated by $P[0]$ is isomorphic to $\mathbf{15}_a$ and thus is complementary to $\mathrm{Sym}^2(M_2(\Gamma_3(2)))$.

5.4.3. The $\mathrm{Sp}(6)$-representation on $M_6(\Gamma_3(2))$

Starting from the basis of the polynomials P_j, it is easy to check by means of a computer that the 680 products $P_i P_j P_k$, $0 \le i \le j \le k \le 14$ are linearly independent in the 870-dimensional vector space $(\mathbb{C}[\ldots, \Theta[\sigma], \ldots]_8)^{H_g}$. The spanned subspace is

$$\mathrm{Sym}^3(M_2(\Gamma_3(2))) := \langle\, P_i P_j P_k \,:\, P_i, P_j, P_k \in M_2(\Gamma_3(2))\,\rangle.$$

The identification $M_2(\Gamma_3(2)) \cong \mathbf{15}_a$, as $\mathrm{Sp}(6,\mathbb{Z}_2)$-representation, implies

$$\mathrm{Sym}^3(M_2(\Gamma_3(2))) \cong 2\cdot\mathbf{15}_a + \mathbf{21}_b + \mathbf{35}_b + \mathbf{84}_a + \mathbf{105}_c + \mathbf{189}_c + \mathbf{216}_a.$$

To decompose all of $V := M_6(\Gamma_3(2))$ we use operator $C = C_V$ as in section 3.8.5.. The eigenvalues λ of C on V and their multiplicities m_λ are easily computed by means of a computer, giving

$$(\lambda, m_\lambda) : (63,1), \quad (27,35), \quad (3,378), \quad (-7,216), \quad (-13,189), \quad (-21,30), \quad (-33,21).$$

$(63,1)$ corresponds to the one-dimensional trivial representation. We would now recognize $\mathrm{Sym}^3(M_2(\Gamma_3(2)))$ in $M_6(\Gamma_3(2))$. From the character table of $\mathrm{Sp}(6)$ in [F1], p.114–115 or in Table 11 (where t_v is in the 16th conjugacy class labeled $1^{-5}2^6$), we see that the irreducible representations ρ of dimension d_λ such that C_ρ has eigenvalue λ in the cases $(\lambda, d_\lambda) = (-13,189), (-33,21)$ are unique. Then, the irreducible representations $\mathbf{189}_c$ and $\mathbf{21}_b$ occur in $M_6(\Gamma_3(2))$, with multiplicity one. The irreducible representations ρ for which C_ρ has eigenvalue $\lambda = 27$ are $\mathbf{21}_a$ and $\mathbf{35}_b$. C has a 35-dimensional eigenspace with $\lambda = 27$, so that the representation $\mathbf{35}_b$ occurs with multiplicity one in $M_6(\Gamma_3(2))$. Also the irreducible representations for which C_ρ has eigenvalue -7 are two: $\mathbf{56}_a$ and $\mathbf{216}_a$. As before, for dimensional reasons, we can conclude that the representation $\mathbf{216}_a$ occurs with multiplicity one in $M_6(\Gamma_3(2))$. The irreducible representations ρ for which C_ρ has eigenvalue $\lambda = 3$ are $\mathbf{105}_c$, $\mathbf{84}_a$, $\mathbf{420}_a$, $\mathbf{210}_b$, then giving rise to two possible decompositions of the 378-dimensional eigenspace with $\lambda = 3$: $2\cdot\mathbf{105}_c + 2\cdot\mathbf{84}_a$ or $\mathbf{210}_b + 2\cdot\mathbf{84}_a$. As $\mathbf{105}_c$ is an irreducible component of $\mathrm{Sym}^3(\mathbf{15}_a)$, it must appear in $M_6(\Gamma_3(2))$, and then $2\cdot\mathbf{105}_c + 2\cdot\mathbf{84}_a$ is the right decomposition. The case when C_ρ has eigenvalue $\lambda = -21$ correspond to the representations: $\mathbf{105}_a$ and $\mathbf{15}_a$. As there is no more space for a 105 dimensional representation, the only possibility is that the 30-dimensional eigenspace of C with $\lambda = -21$ coincides with the representation $2\cdot\mathbf{15}_a$.
We then conclude

$$M_6(\Gamma_3(2)) = \mathrm{Sym}^3(\mathbf{15}_a) + \mathbf{1} + \mathbf{84}_a + \mathbf{105}_c.$$

5.4.4. Asyzygous sextets

In the complement of $\mathrm{Sym}^3(\mathbf{15}_a)$ in $M_6(\Gamma_3(2))$, let us consider the subrepresentation space $\mathbf{105}_c$. This is exactly the same space which has been studied by D'Hoker and Phong in [DP6], as we will see in a moment. Following [DP6], let us consider sets $S = \{\Delta_1, \ldots, \Delta_6\}$ of six totally asyzygous (even) characteristics, which means that $\Delta_i + \Delta_j + \Delta_k$ is odd for distinct i, j, k. An example of such a sextet of even characteristics is

$$S_0 := \{ \left[\begin{smallmatrix}110\\110\end{smallmatrix}\right], \left[\begin{smallmatrix}110\\111\end{smallmatrix}\right], \left[\begin{smallmatrix}111\\110\end{smallmatrix}\right], \left[\begin{smallmatrix}101\\101\end{smallmatrix}\right], \left[\begin{smallmatrix}101\\111\end{smallmatrix}\right], \left[\begin{smallmatrix}111\\101\end{smallmatrix}\right] \}.$$

These are of the form $\left[\begin{smallmatrix}1ab\\1cd\end{smallmatrix}\right]$ where $\left[\begin{smallmatrix}ab\\cd\end{smallmatrix}\right]$ runs over the six odd theta characteristics in genus 2. From the fact that the sum of any three odd characteristics in genus two is even, it follows that the sum of any three of these six characteristics for $g = 3$ is indeed odd. There are 336 asyzygous sextets, on which $\mathrm{Sp}(6)$ acts transitively. The sum of the six characteristics each of these sextets is zero (in $\mathbb{Z}_2$), hence to a sextet S we can associate a modular form F_S of weight 6 on $\Gamma_3(2)$:

$$F_S := \prod_{\Delta \in S} \theta[\Delta]^2 \qquad (\in M_6(\Gamma_3(2))).$$

Each F_S corresponds to a Heisenberg invariant which can be expressed as an homogeneous polynomial of degree 12 in the $\Theta[\sigma]$'s by means of the classical theta formula (54). A computer computation shows that the 336 functions F_S span a 105-dimensional vector space

$$W_{as} := \langle\, F_S \,:\, S \text{ totally asyzygous sextet} \,\rangle \qquad (\subset M_6(\Gamma_3(2))).$$

One can verify indeed that W_{as} has intersection $\{0\}$ with $\mathrm{Sym}^3(M_2(\Gamma_3(2)))$ and that C reduces itself as multiplication by 3 on W_{as}, so that $W_{as} \cong \mathbf{105}_c$, which is what we intended to prove.
In [DP6] it has been shown that W_{as} does not contain functions which transform as $\theta[0]^4$ under $O^+(6)$. This is clear in terms of our analysis, since such a function would generate a subrepresentation of $\mathrm{Ind}_{O^+}^{\mathrm{Sp}}(\epsilon) = \mathbf{15}_a \oplus \mathbf{21}_b$, then contradicting the identification $W_{as} \cong \mathbf{105}_c$.

5.4.5. The $\mathrm{Sp}(6)$-representation on $M_8(\Gamma_3(2))$

The space of modular forms of weight 8 on $\Gamma_3(2)$ is:

$$\mathbb{C}[\ldots, X_\sigma, \ldots]_{16}^{H_g} \cong M_8(\Gamma_3(2)) \oplus \langle F_{16} \rangle,$$

where F_{16} is a homogeneous polynomial of degree 16 in the X_σ such that $F_{16}(\ldots, \Theta[\sigma](\tau), \ldots) = 0$ for all $\tau \in \mathbb{H}_3$, see [vGvdG], [CDG1], and § 4.1:

$$F_{16} = 8 \sum_{\Delta} \theta[\Delta]^{16} - \Big(\sum_{\Delta} \theta[\Delta]^8\Big)^2, \qquad \theta[{}^a_b]^2 = \sum_{\sigma} (-1)^{\sigma b} X_\sigma X_{\sigma+a}$$

so we substitute X_σ rather than $\Theta[\sigma]$ in the classical theta formula. In particular,

$$\dim M_8(\Gamma_3(2)) = 3993 - 1 = 3992.$$

The image $\mathrm{Sym}^4(M_2(\Gamma_3(2)))_0$ of $\mathrm{Sym}^4(M_2(\Gamma_3(2)))$ in this 3992-dimensional vector space is spanned by the products $P_i P_j P_k P_l$, for a basis $P_i \in \mathbb{C}[\ldots, X_\sigma, \ldots]_4^{H_g}$, $1 \le i \le 15$, with 27 independent relations, and includes F_{16}. Thus, its dimension is

$$\dim \mathrm{Sym}^4(M_2(\Gamma_3(2)))_0 = \binom{15+4-1}{4} - 27 - 1 = 3032.$$

Starting from the identification $M_2(\Gamma_3(2)) = \mathbf{15}_a$, a computation with $\mathrm{Sp}(6)$-representations shows that

$$\begin{aligned} \mathrm{Sym}^4(M_2(\Gamma_3(2))) \cong\ & 2\cdot\mathbf{1} + 2\cdot\mathbf{15}_a + \mathbf{27}_a + \mathbf{35}_a + 4\cdot\mathbf{35}_b + 4\cdot\mathbf{84}_a + \mathbf{105}_c + \\ & + \mathbf{168}_a + \mathbf{189}_c + 2\cdot\mathbf{216} + 3\cdot\mathbf{280}_b + \mathbf{336}_a + \mathbf{420}_a. \end{aligned}$$

The $1+27$-dimensional kernel of the map $\mathrm{Sym}^4(M_2(\Gamma_3(2))) \to \mathrm{Sym}^4(M_2(\Gamma_3(2)))_0$ is an $\mathrm{Sp}(6)$-representation, thus it must be $\mathbf{1} + \mathbf{27}_a$.
We then have to identify the complement W in the decomposition

$$M_8(\Gamma_3(2)) \cong \mathrm{Sym}^4(M_2(\Gamma_3(2)))_0 \oplus W, \qquad \dim W = 960.$$

By computing the operator C from section 3.8.5. on $\mathbb{C}[\ldots, X_\sigma, \ldots]_{16}^{H_g}$. we get the pairs given by the eigenvalues λ and their multiplicity m_λ which result to be

$$\begin{array}{llllll} (\lambda, m_\lambda): & (3,1050), & (9,840), & (15,168), & (27,140), & (63,2), \\ & (-3,672), & (-7,648), & (-9,35), & (-13,378), & (-21,60). \end{array}$$

Thus, we see that $\mathbf{27}_a$ cannot be a subrepresentation of $\mathbb{C}[\ldots, X_\sigma, \ldots]_{16}^{H_g}$ because the eigenvalue λ of C on $\mathbf{27}_a$ would have been 35, which is not an eigenvalue of C.
For the remaining eigenvalues we see that:

- the eigenvalue 63 corresponds to the subspace of invariants;
- the eigenvalues $\lambda = 9, -3, -7, -13$ occur only on the irreducible representations $\mathbf{280}_b, \mathbf{336}_a, \mathbf{216}_a, \mathbf{189}_c$ respectively, hence $3 \cdot \mathbf{280}_b + 2 \cdot \mathbf{336}_a + 3 \cdot \mathbf{216}_a + 2 \cdot \mathbf{189}_c$ is a summand of $M_8(\Gamma_3(2))$, and W has a summand $\mathbf{189}_c + \mathbf{216}_a + \mathbf{336}_a$;
- the eigenvalue 3 occurs only on $\mathbf{84}_a$, $\mathbf{105}_c$, $\mathbf{210}_b$ and $\mathbf{420}_a$. Now, $4 \cdot \mathbf{84}_a + \mathbf{105}_c + \mathbf{420}_a$ is a summand of $\mathrm{Sym}^4(M_2(\Gamma_3(2)))_0$, so that there remains a subrepresentation of dimension $1050 - 861 = 189$ in W with the same eigenvalue. Thus $\mathbf{84}_a + \mathbf{105}_c$ is a summand of W;
- the eigenvalue 15 occurs only on $\mathbf{105}_b$, $\mathbf{168}_a$ and $\mathbf{210}_a$. The eigenspace of C for this eigenvalue has dimension 168 so that $\mathbf{168}_a$ is a summand of $M_8(\Gamma_3(2))$ which lies in $\mathrm{Sym}^4(\mathbf{15}_a)$;
- the eigenvalue 27 occurs only on $\mathbf{21}_a$ and $\mathbf{35}_b$. $4 \cdot \mathbf{35}_b$ is a summand of $\mathrm{Sym}^4(M_2(\Gamma_3(2)))_0$ and the dimension of this eigenspace of C is 140. Thus, none of these two representations occurs in W;
- the lowest dimensional representation where the eigenvalue -9 occurs is $\mathbf{35}_a$. As the eigenspace for $\lambda = -9$ has dimension 35, we conclude that $\mathbf{35}_a$ is a summand of $M_8(\Gamma_3(2))$, which lies in $\mathrm{Sym}^4(\mathbf{15}_a)$;
- the eigenvalue -21 occurs only on $\mathbf{15}_a$ and $\mathbf{105}_a$. The eigenspace of C for this eigenvalue has dimension 60, then $4 \cdot \mathbf{15}_a$ is a summand of $M_8(\Gamma_3(2))$ and the summand $2 \cdot \mathbf{15}_a$ is contained in W.

These considerations lead us to the conclusion that

$$W = 2 \cdot \mathbf{15}_a + \mathbf{84}_a + \mathbf{105}_c + \mathbf{189}_c + \mathbf{216}_a + \mathbf{336}_a.$$

5.5. The uniqueness of $\Xi_8[0^{(3)}]$

In section 5.3.5. we have determined a modular form $\Xi_8\left[{000 \atop 000}\right]$ which satisfies the three reduced constraints of section 4..
Such constraints imply that $\Xi_8[0^{(3)}]$ is a modular form on $\Gamma_3(2)$ of weight 8 and should be O^+-invariant (equivalently, it is a modular form on $\Gamma_3(1,2)$), so it must lie in $M_8(\Gamma_3(2))^{O^+}$. As we explained in section 3.8.3., the only $\mathrm{Sp}(6)$-representations which have $O^+(6)$-invariants are $\mathbf{1}$ and $\sigma_\theta = \mathbf{35}_b$. Hence, the decomposition of $M_8(\Gamma_3(2))$ obtained in section 5.4.5. implies that

$$\dim M_8(\Gamma_3(2))^{O^+} = 1+4 = 5.$$

The subspace $M_8(\Gamma_3(2))^{O^+}$ contains the $\mathrm{Sp}(6)$-invariant $\sum_\Delta \theta[\Delta]^{16} \in M_8(\Gamma_3(2))$ as well as the three dimensional subspace

$$\theta\left[{000 \atop 000}\right]^4 M_6(\Gamma_3(2))^\epsilon := \{\theta\left[{000 \atop 000}\right]^4 f : f \in M_6(\Gamma_3(2))^\epsilon\},$$

since both $\theta\left[{000 \atop 000}\right]^4$ and such f are O^+-anti-invariant. A basis of this space is furnished [16], for example, by the three functions $F_i^{(3)}$, with $i = 1,2,3$, of section 5.3.2.. Also we defined the functions $G[\Delta]$ that are modular forms of weight 8 on $\Gamma_3(2)$ and we proved [CDG1] that $G[0] \in M_8(\Gamma_3(2))^{O^+}$.
Using the classical theta formulas, one can check that these functions span the O^+-invariants

$$M_8(\Gamma_4(2))^{O^+} = \theta\left[{000 \atop 000}\right]^4 M_6(\Gamma_4(2))^\epsilon \oplus \langle \sum_\Delta \theta[\Delta]^{16},\ G[0] \rangle.$$

Moreover, the modular form $\Xi_8[0^{(0)}]$ should restrict to $(\theta\left[{0 \atop 0}\right]^4\eta^{12})(\tau_1)(\theta\left[{00 \atop 00}\right]^4\Xi_6\left[{00 \atop 00}\right])(\tau_2)$ on $\mathbb{H}_1 \times \mathbb{H}_2 \subset \mathbb{H}_3$. Note that this restriction is a multiple of $\theta\left[{0 \atop 0}\right](\tau_1)$. Differently from the other four functions, the restriction of $\sum \theta[\Delta]^{16}$ to the diagonal is not a multiple of $\theta\left[{0 \atop 0}\right](\tau_1)$. Thus $\Xi_8[0^{(3)}]$ should be linear combination of just these four functions. The explicit computations done in Section 5.3.5. showed that there is a unique linear combination satisfying the restriction constraint. This verifies the uniqueness of $\Xi_8[0^{(3)}]$.
As we said, as a corollary this implies the non existence of the form $\Xi_6[0^{(3)}]$ proposed by D'Hoker and Phong. It is however interesting to look better at this fact as it gives some new interesting information. The modular form $\Xi_6[0^{(3)}]$ should live in $M_6(\Gamma_g(2))^\epsilon$. Now, the only $\mathrm{Sp}(6)$-representations with O^+-anti-invariants are $\rho_\theta = \mathbf{15}_a$ and $\rho_r = \mathbf{21}_b$, which contain a unique such anti-invariant (cf. 3.8.3.). Thus, from the decomposition of $M_6(\Gamma_3(2))$ given in 5.4.3., it follows that $\dim M_6(\Gamma_3(2))^\epsilon = 3$. One can verify that the following functions are a basis

$$M_6(\Gamma_g(2))^\epsilon = \langle\ \theta\left[{000 \atop 000}\right]^{12},\ \sum_\Delta \theta[\Delta]^{12},\ \theta\left[{000 \atop 000}\right]^4 \sum_\Delta \theta[\Delta]^8\ \rangle.$$

[16] It is clear that a basis for the space $M_6(\Gamma_3(2))^\epsilon$ of $O^+(6)$-anti-invariants is given by $\theta\left[{000 \atop 000}\right]^{12}$, $\theta\left[{000 \atop 000}\right]^4 \sum_\Delta \theta[\Delta]^{12}$ and $\theta\left[{000 \atop 000}\right]^8 \sum_\Delta \theta[\Delta]^8$.

The function $\Xi_6[0^{(3)}]$ should restrict to $\Xi_6[0^{(1)}](\tau_1)\Xi_6[0^{(2)}](\tau_2)$ for $(\tau_1,\tau_2) \in \mathbb{H}_1 \times \mathbb{H}_2 \subset \mathbb{H}_3$, with $\Xi_6[0^{(1)}](\tau_1) = \eta^{12}(\tau_1)$.

The linear combination $a\theta\left[\begin{smallmatrix}000\\000\end{smallmatrix}\right]^{12} + b\sum_\Delta \theta[\Delta]^{12} + c\theta\left[\begin{smallmatrix}000\\000\end{smallmatrix}\right]^4(\sum_\Delta \theta[\Delta]^8)$ restricts, using the results from 5.1. and the restriction of section 5.3.5. (without the common factor $\theta\left[\begin{smallmatrix}000\\000\end{smallmatrix}\right]^4$), to a function of the form $\eta^{12}(\tau_1)g(\tau_2)$ iff

$$a\theta\left[\begin{smallmatrix}00\\00\end{smallmatrix}\right]^{12} + 2b\sum_\delta \theta[\delta]^{12} + 2c\theta\left[\begin{smallmatrix}00\\00\end{smallmatrix}\right]^4 \sum_\delta \theta[\delta]^8 = 0$$

on $\mathbb{H}_2$. However, it follows from the results in Table 5.2.1. that the ten even $\theta[\delta]^{12}$'s span a ten dimensional space which does not contains $\theta\left[\begin{smallmatrix}00\\00\end{smallmatrix}\right]^4 \sum_\delta \theta[\delta]^8$. Hence we must have $a = b = c = 0$, which proves that there is no function $\Xi_6[0^{(3)}]$ satisfying the three constraints imposed in [DP6].

5.6. The case $g = 4$

Finally, we will consider here the case of genus four, where we will again be able to find a suitable modular form Ξ_8 satisfying the four constraints of section 2.. Thus, this function turn out to be a good candidate for the superstring measure, even though the uniqueness is true in a weaker form w.r.t. the previous cases. The construction procedure is the same as in the previous cases: we search functions that are $O^=$-invariant and then look for a linear combination satisfying the factorization constraint. However, in this case we make use of a further assumption: we require for the forms $\Xi_8[\Delta]$ to be polynomials in the theta constants. This new assumption is motivated by the non normality of the ring of modular forms in genus four, thus it could contain modular forms that are not polynomial in theta constants. The uniqueness (in this sense) will follow, as before, from the fact we use a basis for the space of (polynomial) modular forms on $\Gamma_4(1,2)$, but is then restricted to the polynomial subspace. The factorization constraints which must be imposed to the forms Ξ_8 when the period matrix $\tau_4 \in \mathbb{H}_4$ becomes reducible, are $\Xi_8[0^{(4)}](\tau_{k,4-k}) = \Xi_8[0^{(k)}](\tau_k)\Xi_8[0^{(4-k)}](\tau_{4-k})$, with $\tau_{k,4-k} := \left(\begin{smallmatrix}\tau_k & 0\\ 0 & \tau_{4-k}\end{smallmatrix}\right) \in \mathbb{H}_4$. As in genus three, first we construct the forms Ξ_8, then we tackle the problem of the uniqueness.

5.6.1. The modular forms $F_i^{(4)}$, $G_1^{(4)}[0^{(4)}]$ and $G_2^{(4)}[0^{(4)}]$

It is easy to write down some of the O^+-invariants in genus four as extensions of the functions defined in genus two and three. An obvious generalization of the functions F_i introduced in sections 5.2.2.and 5.3.2. is:

$$F_1^{(4)} := \theta\left[\begin{smallmatrix}0000\\0000\end{smallmatrix}\right]^{16}, \qquad F_2^{(4)} := \theta\left[\begin{smallmatrix}0000\\0000\end{smallmatrix}\right]^4 \sum_\Delta \theta[\Delta]^{12}, \qquad F_3^{(4)} := \theta\left[\begin{smallmatrix}0000\\0000\end{smallmatrix}\right]^8 \sum_\Delta \theta[\Delta]^8,$$

where the sum is over the 136 even characteristics Δ in genus four. These functions are modular forms of weight 8 on $\Gamma_4(1,2)$. When clear from the contest we omit the apex indicating the genus. The restriction of these forms to $\mathbb{H}_1 \times \mathbb{H}_3$ and $\mathbb{H}_2 \times \mathbb{H}_2$ are

$$
\begin{aligned}
F_1^{(4)}(\tau_{2,2}) &= \theta\left[\begin{smallmatrix}00\\00\end{smallmatrix}\right]^{16}(\tau_2)\theta\left[\begin{smallmatrix}00\\00\end{smallmatrix}\right]^{16}(\tau_2') \\
&= F_1^{(2)}(\tau_2)F_1^{(2)}(\tau_2'), \\
F_2^{(4)}(\tau_{2,2}) &= \left(\theta\left[\begin{smallmatrix}00\\00\end{smallmatrix}\right]^4 \textstyle\sum_\delta \theta[\delta]^{12}\right)(\tau_2)\left(\theta\left[\begin{smallmatrix}00\\00\end{smallmatrix}\right]^4 \textstyle\sum_\delta \theta[\delta]^{12}\right)(\tau_2') \\
&= F_2^{(2)}(\tau_2)F_2^{(2)}(\tau_2'), \\
F_3^{(4)}(\tau_{2,2}) &= \theta\left[\begin{smallmatrix}00\\00\end{smallmatrix}\right]^8(\tau_2)\theta\left[\begin{smallmatrix}00\\00\end{smallmatrix}\right]^8(\tau_2')\left(\textstyle\sum_\delta \theta[\delta]^8\right)(\tau_2)\left(\textstyle\sum_\delta \theta[\delta]^8\right)(\tau_2') \\
&= F_3^{(2)}(\tau_2)F_3^{(2)}(\tau_2'),
\end{aligned}
$$

whereas the restrictions to $\mathbb{H}_1 \times \mathbb{H}_3$ are

$$F_1^{(4)}(\tau_{1,3}) = \left(\theta\left[\begin{smallmatrix}0\\0\end{smallmatrix}\right]^4(\tfrac{1}{3}f_{21} + \eta^{12})\right)(\tau_1)F_1^{(3)},(\tau_3)$$

$$F_2^{(4)}(\tau_{1,3}) = \left(\theta\left[\begin{smallmatrix}0\\0\end{smallmatrix}\right]^4(\tfrac{2}{3}f_{21} - \eta^{12})\right)(\tau_1)F_2^{(3)}(\tau_3),$$

$$F_3^{(4)}(\tau_{1,3}) = \left(\theta\left[\begin{smallmatrix}0\\0\end{smallmatrix}\right]^4\tfrac{2}{3}f_{21}\right)(\tau_1)F_3^{(3)}(\tau_3).$$

The modular form $G^{(3)}$ of section 5.3.4. can be generalized in two ways. The first one is to stick to three dimensional isotropic subspaces W in $\mathbb{Z}_2^8$. Given such a W, there are $3 \cdot 8 = 24$ even quadrics Q_Δ such that $W \subset Q_\Delta$. Let $Q_0 \subset \mathbb{Z}_2^8$ be the even quadric with characteristic $\Delta_0 = [0^{(4)}]$. We can use the octets of quadrics which contain Q_0 to define a modular form $G_1[0]$:

$$G_1\left[\begin{smallmatrix}0000\\0000\end{smallmatrix}\right] := \sum_{W \subset Q_0} \prod_{w \in W} \theta[\Delta_0 + w]^2,$$

where the sum is extended to all the 2025 three dimensional isotropic subspaces $W \subset Q_0$, and for each such subspace we take the product of the eight even $\theta[\Delta_0 + w]^2$. As $0 \in W$, for any subspace W, the function $G_1[0]$ is a multiple of $\theta[\Delta_0]^2$. The function $G_1[0]$ is a modular form on $\Gamma_4(1,2)$ of weight 8, as can be shown using the explicit transformation theory of theta functions as in Section 3.6. for genus three or in the Appendices of [CDG1] (or cf. [I2] or see [Gr], Proposition 13).
Using methods similar to those in Appendix C of [CDG1] for the case $g = 3$ we find the restriction of $G_1[0]$ to $\mathbb{H}_1 \times \mathbb{H}_3$:

$$
\begin{aligned}
G_1\left[\begin{smallmatrix}0000\\0000\end{smallmatrix}\right](\tau_{1,3}) &= \theta\left[\begin{smallmatrix}0\\0\end{smallmatrix}\right]^{16}(\tau_1)G\left[\begin{smallmatrix}000\\000\end{smallmatrix}\right](\tau_3) + \left(\theta\left[\begin{smallmatrix}0\\0\end{smallmatrix}\right]^8(\theta\left[\begin{smallmatrix}0\\1\end{smallmatrix}\right]^8 + \theta\left[\begin{smallmatrix}1\\0\end{smallmatrix}\right]^8)\right)(\tau_1)\left(H\left[\begin{smallmatrix}000\\000\end{smallmatrix}\right] + 7G\left[\begin{smallmatrix}000\\000\end{smallmatrix}\right]\right)(\tau_3) \\
&= \theta\left[\begin{smallmatrix}0\\0\end{smallmatrix}\right]^4(\tau_1)\left(\tfrac{1}{3}f_{21}(\tau_1)(H\left[\begin{smallmatrix}000\\000\end{smallmatrix}\right] + 8G\left[\begin{smallmatrix}000\\000\end{smallmatrix}\right])(\tau_3) - \eta^{12}(\tau_1)(H\left[\begin{smallmatrix}000\\000\end{smallmatrix}\right] + 6G\left[\begin{smallmatrix}000\\000\end{smallmatrix}\right])(\tau_3\right)
\end{aligned}
$$

where

$$H\left[\begin{smallmatrix}000\\000\end{smallmatrix}\right] := \sum_{W' \subset Q_0} \prod_{w \in W} \theta[\Delta_0^{(3)} + w]^4,$$

and f_{21} as in Section 5.1.. The sum in H is over the 105 isotropic 2-dimensional subspaces W' contained in Q_0, where now $Q_0 \subset \mathbb{Z}_2^3$.
Similarly, the restriction of $G_1[0]$ to $\mathbb{H}_2 \times \mathbb{H}_2$ is

$$\begin{aligned} G_1\left[\begin{smallmatrix}0000\\0000\end{smallmatrix}\right](\tau_{2,2}) &= \theta\left[\begin{smallmatrix}00\\00\end{smallmatrix}\right]^4(\tau_2)g(\tau_2)F_3^{(2)}(\tau_2') + F_3^{(2)}(\tau_2)\theta\left[\begin{smallmatrix}00\\00\end{smallmatrix}\right]^4(\tau_2')g(\tau_2') \\ &\quad + 9\theta\left[\begin{smallmatrix}00\\00\end{smallmatrix}\right]^4(\tau_2)g(\tau_2)\theta\left[\begin{smallmatrix}00\\00\end{smallmatrix}\right]^4(\tau_2')g(\tau_2'), \end{aligned}$$

with $F_3^{(2)}$ as in section 5.2.2.,

$$g(\tau_2) = \sum_{W'\subset Q_0} \prod_{w\in W'-\{0\}} \theta[\Delta_0^{(2)} + w]^4(\tau_2),$$

where the sum is over the 6 isotropic 2-dimensional subspaces W' of $Q_0 \subset \mathbb{Z}_2^4$, and we take a product of only three terms (the factor $\theta[0^{(2)}]^4$ for $w = 0$ is taken out in the formula for $G_1[0](\tau_{2,2})$).
Another generalization of G makes use of Lagrangian subspaces $L \cong \mathbb{Z}_2^4$ of $V = \mathbb{Z}_2^8$. For each L there are 16 even quadrics Q_Δ with $L \subset Q_\Delta$. For an even characteristic Δ we define

$$G_2[\Delta] = \sum_{L\subset Q_\Delta} \prod_{Q\supset L} \theta[\Delta_Q],$$

the sum being over the 270 Lagrangian subspaces L of $V = \mathbb{Z}_2^8$ which are contained in Q_Δ. Again, $L \subset Q_\Delta$ implies that $G_2[\Delta]$ is a multiple of $\theta[\Delta]$. The function $G_2[0]$ is also modular form on $\Gamma_4(1,2)$ of weight 8.
The restriction of $G_2[0]$ to $\mathbb{H}_1 \times \mathbb{H}_3$ is

$$\begin{aligned} G_2\left[\begin{smallmatrix}0000\\0000\end{smallmatrix}\right](\tau_{1,3}) &= \left(\theta\left[\begin{smallmatrix}0\\0\end{smallmatrix}\right]^8(\theta\left[\begin{smallmatrix}0\\1\end{smallmatrix}\right]^8 + \theta\left[\begin{smallmatrix}1\\0\end{smallmatrix}\right]^8)\right)(\tau_1)G\left[\begin{smallmatrix}000\\000\end{smallmatrix}\right](\tau_3) \\ &= \left(\theta\left[\begin{smallmatrix}0\\0\end{smallmatrix}\right]^4(\tfrac{1}{3}f_{21} - \eta^{12})\right)(\tau_1)G\left[\begin{smallmatrix}000\\000\end{smallmatrix}\right](\tau_3), \end{aligned}$$

whereas its to $\mathbb{H}_2 \times \mathbb{H}_2$ is as follows:

$$G_2\left[\begin{smallmatrix}0000\\0000\end{smallmatrix}\right](\tau_{2,2}) = \theta\left[\begin{smallmatrix}00\\00\end{smallmatrix}\right]^4(\tau_2)\theta\left[\begin{smallmatrix}00\\00\end{smallmatrix}\right]^4(\tau_2')g(\tau_2)g(\tau_2'),$$

with g as above.
Before considering the restriction to $\mathbb{H}_1 \times \mathbb{H}_3$ of the form $\Xi_8^{(4)}$ let us give a couple of identities which we need to express the restrictions of the $G_i^{(4)}[0]$'s as linear combinations of the $F_i^{(3)}$'s and $G^{(3)}[\Delta]$ in genus three and the $F_i^{(2)}$'s in genus two:

$$H\left[\begin{smallmatrix}000\\000\end{smallmatrix}\right] = (2F_1^{(3)} + 8F_2^{(3)} - 3F_3^{(3)})/6, \qquad \theta\left[\begin{smallmatrix}00\\00\end{smallmatrix}\right]^4 g = (2F_1^{(2)} + 4F_2^{(2)} - 3F_3^{(2)})/6. \quad (102)$$

These identities can be verified using the classical theta formula (it is helpful to use a computer as well).
Let us now consider a general linear combination of the 5 functions

$$\Xi_8\left[\begin{smallmatrix}0000\\0000\end{smallmatrix}\right] = a_1F_1^{(4)} + \ldots + a_4G_2[0] + a_5G_1[0],$$

and try first to impose the restriction to $\mathbb{H}_1 \times \mathbb{H}_3$:

$$\Xi_8\left[\begin{smallmatrix}0000\\0000\end{smallmatrix}\right](\tau_{1,3}) = \Xi_8\left[\begin{smallmatrix}0\\0\end{smallmatrix}\right](\tau_1)\Xi_8\left[\begin{smallmatrix}000\\000\end{smallmatrix}\right](\tau_3) = \left(\theta\left[\begin{smallmatrix}0\\0\end{smallmatrix}\right]^4\eta^{12}\right)(\tau_1)\Xi_8\left[\begin{smallmatrix}000\\000\end{smallmatrix}\right](\tau_3). \qquad (103)$$

As function of $\tau_1 \in \mathbb{H}_1$, this restriction is a linear combination of $\Xi_8[0^{(1)}]$ and $\theta[0^{(1)}]^4 f_{21}$. The second type terms however must disappear in the correct factorization so that $\Xi_8\left[\begin{smallmatrix}0000\\0000\end{smallmatrix}\right](\tau_{1,3})$ is a multiple of $\eta^{12}(\tau_1)$ iff

$$a_1F_1 + 2a_2F_2 + 2a_3F_3 + a_4G\left[\begin{smallmatrix}000\\000\end{smallmatrix}\right] + a_5(H\left[\begin{smallmatrix}000\\000\end{smallmatrix}\right] + 8G\left[\begin{smallmatrix}000\\000\end{smallmatrix}\right]) = 0.$$

Using the formula for H given in (102) we get:

$$(a_1 + \frac{1}{3}a_5)F_1 + (2a_2 + \frac{4}{3}a_5)F_2 + (2a_3 - \frac{1}{2}a_5)F_3 + (a_4 + 8a_5)G\left[\begin{smallmatrix}000\\000\end{smallmatrix}\right] = 0.$$

As the four functions here are independent (cf. [DvG]), we get the solutions

$$(a_1, a_2, a_3, a_4, a_5) = \lambda(-2, -4, 3/2, -48, 6), \qquad (\lambda \in \mathbb{C}).$$

For such a_i the linear combination $a_1\theta[0^{(4)}]^4F_1 + \ldots + a_4G_2[0] + a_5G_1[0]$ restricts to:

$$\theta\left[\begin{smallmatrix}0\\0\end{smallmatrix}\right]^4\eta^{12}\left(a_1F_1 - a_2F_2 - a_4G\left[\begin{smallmatrix}000\\000\end{smallmatrix}\right] - a_5(H\left[\begin{smallmatrix}000\\000\end{smallmatrix}\right] + 6G\left[\begin{smallmatrix}000\\000\end{smallmatrix}\right])\right)$$

which, using again the formula for H gives a genus three factor

$$(a_1 - \frac{1}{3}a_5)F_1 - (a_2 + \frac{4}{3}a_5) + \frac{1}{2}a_5F_3 - (a_4 + 6a_5)G\left[\begin{smallmatrix}000\\000\end{smallmatrix}\right].$$

Setting $\lambda = -2$, so that $(a_1, \ldots, a_5) = (4, 8, -3, 96, -12)$, we get (cf. 5.3.5.)

$$8F_1 + 8F_2 - 6F_3 - 24G\left[\begin{smallmatrix}000\\000\end{smallmatrix}\right] = 24\Xi_8\left[\begin{smallmatrix}000\\000\end{smallmatrix}\right].$$

Thus, the function

$$\Xi_8\left[\begin{smallmatrix}0000\\0000\end{smallmatrix}\right] := \left(4F_1 + 8F_2 - 3F_3 + 96G_2\left[\begin{smallmatrix}0000\\0000\end{smallmatrix}\right] - 12G_1\left[\begin{smallmatrix}0000\\0000\end{smallmatrix}\right]\right)/24 \qquad (104)$$

satisfies correctly the constraint on the restriction to $\mathbb{H}_1 \times \mathbb{H}_3$.
However, having found the solution, we must check that it satisfy correctly the restriction to $\mathbb{H}_2 \times \mathbb{H}_2$ also. It is useful to observe that:

$$\begin{aligned} 24\Xi_8\left[\begin{smallmatrix}0000\\0000\end{smallmatrix}\right](\tau_{2,2}) &= \left(4F_1 + 8F_2 - 3F_3 + 96G_2 - 12G_1\right)(\tau_{2,2}) \\ &= \theta\left[\begin{smallmatrix}00\\00\end{smallmatrix}\right]^4(\tau_2)\theta\left[\begin{smallmatrix}00\\00\end{smallmatrix}\right]^4(\tau_2')h(\tau_2, \tau_2'), \end{aligned}$$

with h a holomorphic function which we will not write down explicitly. Taking out the factor $\theta[0^{(2)}]^4(\tau_2)\theta[0^{(2)}]^4(\tau_2')$, which also occurs in $\Xi_8[0^{(2)}](\tau_2)\Xi_8[0^{(2)}](\tau_2')$, simplifies the computation. One finds, using the classical theta formula and a computer, that $h(\tau_2, \tau_2') = \Xi_6[0^{(2)}](\tau_2)\Xi_6[0^{(2)}](\tau_2')$ and thus:

$$\Xi_8\left[\begin{smallmatrix}0000\\0000\end{smallmatrix}\right](\tau_{2,2}) = \Xi_8\left[\begin{smallmatrix}00\\00\end{smallmatrix}\right](\tau_2)\Xi_8\left[\begin{smallmatrix}00\\00\end{smallmatrix}\right](\tau_2').$$

Therefore the modular form $\Xi_8[0^{(4)}]$ on $\Gamma_4(1,2)$ of weight 8, defined in (104), satisfies all the factorization constraints in genus four. Using the representation theory of group we will be also able to prove the uniqueness of this modular form.

5.7. The $\mathrm{Sp}(8)$-representation on $M^\theta_{2k}(\Gamma_4(2))$

In case $g = 4$ we are no longer sure if $M^\theta_{2k}(\Gamma_4(2))$, the space of modular forms of weight of $2k$ which are (Heisenberg-invariant) polynomials in the $\Theta[\sigma]$'s, is equal to all of $M_{2k}(\Gamma_4(2))$ (cf. [OSM]).
The $\mathrm{Sp}(8)$-representation on $M^\theta_2(\Gamma_4(2))$, which we denoted by ρ_θ in section 3.8.3., is the unique 51-dimensional irreducible representation of $\mathrm{Sp}(8)$ (a table of the 81 irreducible representations of $\mathrm{Sp}(8)$ can be easily generated with the computer algebra program 'Magma'). The complement of $\mathrm{Sym}^2(M^\theta_2(\Gamma_4(2))$ in $M^\theta_4(\Gamma_4(2))$ (of $\mathrm{Sp}(8)$-representations) now has codimension 918:

$$\dim M^\theta_4(\Gamma_4(2)) - \dim \mathrm{Sym}^2(M^\theta_2(\Gamma_4(2))) = 2244 - \binom{51+1}{2} = 2244 - 1326 = 918.$$

We computed the operator C from section 3.8.5. on $M^\theta_4(\Gamma_4(2))$. The resulting pairs of its eigenvalues λ with multiplicity m_λ are:

$$(\lambda, m_\lambda): \qquad (-25, 918), \quad (39, 1190), \quad (119, 135), \quad (255, 1).$$

The last three eigenspaces of C correspond to the irreducible representations σ_c, σ_θ and $\mathbf{1}$ respectively and their direct sum is $\mathrm{Sym}^2(\rho_\theta)$, cf. 3.8.4.. The character table shows that there are only 10 irreducible representations of $\mathrm{Sp}(8)$ with dimension less then 918 and there is a unique irreducible representation with dimension 918. However, of these eleven irreducible representations, the map C has eigenvalue $\lambda = -25$ only on the one of dimension 918. Thus we conclude that $M^\theta_4(\Gamma_4(2))$ is the sum of just four irreducible representations (like $M^\theta_4(\Gamma_3(2))$, cf. section 5.4.2.). In principle, in this way, one could decompose also the representations on the space of modular forms of weight six and eight, as in the case of $\mathrm{Sp}(6)$, but the computation of the Casimir C_ρ is very time and memory consuming.

5.7.1. The uniqueness of $\Xi_8[0^{(4)}]$

Using the same methods as in section 5.5. for the case $g = 3$ and the observation in section 5.2.2. for the case $g = 2$,, we can show that the three constraints characterize the form $\Xi_8[0^{(4)}]$ up to an additive term λJ, where $J = 0$ defines the Jacobi locus J_4 (the locus of period matrices of Riemann surfaces in $\mathbb{H}_4$) and $\lambda \in \mathbb{C}$. In fact, J is a modular form of weight 8 on Γ_4, so it can be added to $\Xi_8[0^{(4)}]$ without changing its $\Gamma_g(1,2)$-invariance. Moreover $\mathbb{H}_1 \times \mathbb{H}_3$ and $\mathbb{H}_2 \times \mathbb{H}_2$ are contained in the closure of the Jacobi locus, so J is zero on these loci.
The key point is the determination of the dimension of $M^\theta_8(\Gamma_4(2))^{O^+}$, which could be done by computer. M. Oura determined this dimension using the methods from [R1], [R2]: $\dim M^\theta_8(\Gamma_4(2))^{O^+} = 7$. We already know 7 independent functions in $M^\theta_8(\Gamma_4(2))^{O^+}$. It follows that $M_8(\Gamma_4(2))^{O^+} = M^\theta_8(\Gamma_4(2))^{O^+}$ and that

$$M_8(\Gamma_4(2))^{O^+} = \langle \sum_\Delta \theta[\Delta]^{16},\ (\sum_\Delta \theta[\Delta]^8)^2,\ F_1,\ F_2,\ F_3,\ G_1[0^{(4)}],\ G_2[0^{(4)}] \rangle.$$

The modular form J vanishing on the Jacobi locus is

$$J = 16\sum\theta[\Delta]^{16} - (\sum\theta[\Delta]^{8})^{2}$$

(cf. [I3]). Now the proof of the uniqueness of $\Xi_8[0^{(4)}]$ in $M_8^{\theta}(\Gamma_g(2))$ can be obtained with arguments similar to those in sections 5.5. and 5.2.2..

5.8. The cosmological constant in $g = 3$, $g = 4$

Supersymmetry impose the same number of fermionic and bosonic states. This leads to the non renormalization theorems [Ma1, Ma2, Mo2] which in particular require for the zero points function, t.i. the cosmological constant, to vanishes identically. We will prove that our solution for the chiral superstring measure gives a vanishing contribution to the cosmological constant for both the case $g = 3$ and $g = 4$. Independent proofs of this result in $g = 4$, which make use of different techniques, can be found in [Gr] or in [SM2]. The GSO projection for type II superstring gives for the phases in (15) the value $c_{\Delta,\Delta'} = 1$ so that we will prove that $\sum_{\Delta} \mathrm{d}\mu[\Delta] = 0$ or, equivalently, $\left(\sum_{\Delta}\Xi_8[\Delta]\right)(\Omega) = 0$.

5.8.1. The case $g = 3$

The sum of the 36[17] functions $\Xi_8[\Delta]$ is invariant under $\mathrm{Sp}(6)$, hence it is a modular form of weight 8 on $Sp(6,\mathbb{Z})$. In the decomposition of $M_8(\Gamma_3(2))$, see section 5.4.5., the representation **1** has multiplicity one, thus there is a unique, up to a scalar multiple, $\mathrm{Sp}(6)$-invariant on $M_8(\Gamma_3(2))$. This invariant is $\sum_{\Delta}\theta[\Delta]^{16}$. Hence, the sum of the functions $\Xi_8[\Delta]$ must be a scalar multiple of this invariant:

$$\Big(\sum_{\Delta}\Xi_8[\Delta]\Big)(\tau) = \mu\Big(\sum_{\Delta}\theta[\Delta]^{16}\Big)(\tau). \tag{105}$$

Note that the function $\sum_{\Delta}\Xi_8[\Delta]$, obtained from the $\Xi_8[0]$ of section 5.3.5. with the transformation formula of the theta constants, is given by

$$-4\sum_{\Delta}\theta[\Delta]^{16} - 4\sum_{\Delta}\theta[\Delta]^{4}\Big(\sum_{\Delta'}\epsilon_{\Delta,\Delta'}\theta[\Delta']^{12}\Big) + 3\Big(\sum_{\Delta}\theta[\Delta]^{8}\Big)^{2} + 12\sum_{\Delta}G[\Delta],$$

where, again, the constants $\epsilon_{\Delta,\Delta'} = \pm 1$ are determined by the transformation theory. We will now show that $\mu = 0$ by looking first at diagonal form period matrices $\tau = \mathrm{diag}(\tau_1,\tau_2,\tau_3)$ and then setting $\tau_1,\tau_2,\tau_3 \to i\infty$. On the theta constants this gives

$$\theta\left[\begin{smallmatrix}abc\\def\end{smallmatrix}\right] \longmapsto \begin{cases} 1 & \text{if } a=b=c=0,\\ 0 & \text{else,}\end{cases}$$

hence $\sum_{\Delta}\theta[\Delta]^{16} \mapsto 8$ and $\sum_{\Delta}\theta[\Delta]^{8} \mapsto 8$. In the summand $\sum_{\Delta}\theta[\Delta]^{4}(\sum_{\Delta'}\epsilon_{\Delta,\Delta'}\theta[\Delta']^{12})$ we thus only need to consider the terms with[18] $\Delta = \left[\begin{smallmatrix}0\\b\end{smallmatrix}\right]$,

[17] This is the number of the even characteristics in $g = 3$.

[18] Here we mean $0 = (0,0,0)$ and $b = (b_1,b_2,b_3)$

$\Delta' = [{0 \atop b'}]$. The terms with $\Delta = [{0 \atop b}]$ are summands of $\Xi_8[{0 \atop b}](\tau)$. Let M be the symplectic matrix

$$M = \begin{pmatrix} I & B \\ 0 & I \end{pmatrix}, \qquad B = \mathrm{diag}(b_1, b_2, b_3), \qquad \text{so} \quad M \cdot [{0 \atop b'}] = [{0 \atop b+b'}].$$

In particular, $M \cdot [{0 \atop 0}] = [{0 \atop b}]$ and thus $\Xi_8[{0 \atop b}](\tau) = \Xi_8[{0 \atop 0}](M^{-1}\tau)$ (note that $\gamma(M, M^{-1} \cdot \tau) = 1$). From the definition of the theta constants as series in 3.4. it is obvious that

$$\theta[{0 \atop b'}]^4(M^{-1} \cdot \tau) \;=\; \theta[{0 \atop b+b'}]^4(\tau)$$

hence $\epsilon_{\Delta,\Delta'} = +1$ if $\Delta = [{0 \atop b}]$, $\Delta' = [{0 \atop b'}]$. Thus we get

$$\sum_{\Delta} \theta[\Delta]^4 (\sum_{\Delta'} \epsilon_{\Delta,\Delta'} \theta[\Delta']^{12}) \longmapsto \sum_{b} \theta[{0 \atop b}]^4 (\sum_{b'} \epsilon_{[{0 \atop b}],[{0 \atop b'}]} \theta[{0 \atop b'}]^{12}) \longmapsto 8 \cdot 8 = 64.$$

Finally, each $G[\Delta]$ is a sum of P_L's and each P_L is a product of eight distinct theta constants. Thus all P_L's map to zero except for $P_{L_0} := \prod_{d,e,f} \theta[{000 \atop def}]$ which maps to 1. Note that $L_0 = \{ ({abc \atop 000}) \}$ and that $L_0 \subset Q_\Delta$ iff $\Delta = [{000 \atop def}]$. Thus exactly 8 of the $G[\Delta]$ map to one, and the others map to zero. The constant μ can now be determined

$$-4 \cdot 8 - 4 \cdot 8^2 + 3 \cdot 64 + 12 \cdot 8 = \mu \cdot 8 \qquad \Longrightarrow \qquad \mu = 0,$$

hence the cosmological constant is zero.

5.8.2. The case $g = 4$

In this case we have not the full decomposition in irreducible of the space $M_8(\Gamma_4(2))$. Nevertheless, we are able to determine, using a computer, that the dimension of the space of Sp(8)-invariant in $\mathbb{C}[\cdots, X_\sigma, \cdots]^{H_g}]$ is two. The space $M_8(\Gamma_4(2))$ is huge, so one starts with finding invariants for the transvections which acts 'diagonally' on the X_σ to reduce the computation to a smaller space. These two invariants correspond to the modular forms Ψ_8 and Ψ_4^2, with $\Psi_{4k}(\tau) := \sum_\Delta \theta[\Delta^{(4)}]^{8k}(\tau)$. The combination $J = 16\Psi_8 - \Psi_4^2$ vanishes on the Jacobi locus, i.e. the space of the matrices $\tau \in H_g$ that are, also, a period matrix of some Riemann surfaces. In genus four there are 136 even characteristics, thus we have to consider the sum of the 136 functions $\Xi_8[\Delta]$. As $\sum_\Delta \Xi_8[\Delta^{(4)}]$ is an $Sp(8)$-invariant of weight 8, there are constants λ, μ such that

$$\sum_{\Delta} \Xi_8[\Delta^{(4)}] \;=\; \lambda \Psi_8 \;+\; \mu \Psi_4^2 \tag{106}$$

and it suffices to show that $\lambda + 16\mu = 0$. For this one can specialize τ to a diagonal matrix $\tau = \mathrm{diag}(\tau_1, \ldots, \tau_4)$ and then let $\tau_j \mapsto \infty$ for $j = 1, \ldots, 4$, similar to the computation in genus three of the previous section.

6. Conclusion

We have provided a detailed explanation of the results and the methods adopted in [CD, CDG1, CDG2] to determine a good candidate for the superstring measure. Our strategy, yet adopted by D'Hoker and Phong [DP5, DP6] and by many other authors before (see [Mo] and references therein), is not to provide a direct computation from first principles, but consists in looking for reasonable ansätze (surely inspired by first principles) which should then lead to a unique solution which must then satisfy a number of tests. As we seen, the ansatz we have chosen is very closed to the one of D'Hoker and Phong, but the very slight modification has been proved to be crucial in providing a unique solution. This ansatz is the more general one after the assumption of the validity of the relation (11). We seen that such assumption is highly criticizable, but the fact that it gives rise to the existence of a unique solution is quite encouraging, especially at genus $g \leq 3$. For genus four, the result is weakened by the fact that the Shottky set has strict positive codimension in the Siegel upper half plane, and that, indeed, uniqueness comes out to be proved in a restricted form. We have also seen that, as predicted by supersymmetry consistence, the cosmological constant is predicted to vanish by our solutions. In [Mo1] it has been shown that also the two-point and the three-point functions vanish, according to the non renormalization theorems ([Ma1, Ma2, Mo2]), but the proof is provided for hyperelliptic surfaces, which are a zero measure subset for genus higher than 2. A complete proof of the vanishing of the two-point function at genus $g = 3$ for our solution has been provided in [GSM]. The check of the same condition for the three point function and for the $g = 4$ case have yet to be provided beyond the hyperelliptic case.
It is indispensable to mention that, after [CDG1], other papers appeared providing new expressions and generalizations of our results, see for example [Gr, SM1, MV, OPSMY]. In particular, the remarkable paper of Grushevsky [Gr] provided an elegant formal expression for the solution at any genus g. Unfortunately, such expressions involve square and higher order roots of modular forms, which are not well defined in general. In [SM1] it has been proved that the Grushevsky's expression is well defined for $g = 5$. Another candidate for the $g = 5$ case is determined in [OPSMY]. It is not yet clear if the two expressions are indeed equivalent. In any case, as remarked in [DBMS] the solutions at genus 5 are no more uniquely determined by the ansatz and the vanishing of the cosmological constant is indeed added as a further condition. This means that the ansatz does not definitively encode the whole physical requirements and one is led to turn back to a direct analysis by first principles, as done by D'Hoker and Phong for the genus two case.
Finally, even though the expressions we have found provide explicit results, some of the modular forms used for the construction need to be recast in some easier form to become manageable for concrete applications.

Acknowledgments

We are strongly indebted to Bert Van Geemen, from who we have learned all the mathematical tools used to reach our goals and whose collaboration has been and continue to be fundamental. We thank Gilberto Bini, Matteo Cardella and Riccardo Salvati-Manni for useful discussions. S.L.C. is grateful to Frank Columbus for the invitation to write this contribution.

A Restricted and Induced Representations

In this section we introduce restricted and induced representations and we will see as they characters are connected by the Frobenius reciprocity [S]. Let G be a group and H a subgroup. Suppose ρ is a matrix representation of G, then one can obtain a representation of H by the operation of restriction. The restriction of ρ to H, $\mathrm{Res}_H^G(\rho)$, is given by:

$$\mathrm{Res}_H^G(\rho(h)) := \rho(h), \tag{107}$$

for all $h \in H$. The restriction, actually, is a representation of H. Although ρ may be an irreducible representation of G, $\mathrm{Res}_H^G(\rho(h))$ can be reducible. We can also consider the inverse operation. The process of moving from a representation of the subgroup H to a representation of the whole G is called induction. Fix a transversal $\{t_1, \cdots, t_k\}$ for the left cosets of H, i.e. $\mathcal{H} = \{t_1H, \cdots, t_kH\}$ is a complete set of disjoint left cosets for H in G, so $G = t_1H \uplus \cdots \uplus t_kH$ and $\uplus$ denotes disjoint union, $t_i \in G$. Let σ be a representation of H, then the induced representation $\mathrm{Ind}_H^G(\sigma)$ from H to G assigns to each $g \in G$ the block matrix:

$$\mathrm{Ind}_H^G(\sigma(g)) := \sigma(t_i^{-1}gt_j) = \begin{pmatrix} \sigma(t_1^{-1}gt_1) & \cdots & \sigma(t_1^{-1}gt_k) \\ \vdots & \ddots & \vdots \\ \sigma(t_k^{-1}gt_1) & \cdots & \sigma(t_k^{-1}gt_k), \end{pmatrix} \tag{108}$$

and $\sigma(g)$ is the zero matrix if $g \notin H$. It can be proved that $\mathrm{Ind}_H^G(\sigma)$ is a representation of G.

Of particular interest is the induced representation of the identity $\mathbf{1}$ and it is strictly related to the coset representation. Suppose, as before, that $\mathcal{H} = \{t_1H, \cdots, t_kH\}$ is a complete set of disjoint left cosets for H in G. The group G acts on the set $\mathcal{H}$ by $g(g_iH) := (gg_i)H$, for all $g \in G$. The set $\mathcal{H}$ can be turned, as every set on which a group G acts, in a G-module as follows. Let $\mathbb{C}\mathcal{H}$ denote the vector space generated by $\mathcal{H}$ over $\mathbb{C}$, that is $\mathcal{H}$ consists of all the formal linear combinations $a_1g_1H + \cdots + a_kg_kH$, $a_i \in \mathbb{C}$. Vector addition and scalar multiplication are defined as follows:

$$\begin{aligned} &(a_1g_1H + \cdots + a_kg_kH) + (b_1g_1H + \cdots + b_kg_kH) = (a_1 + b_1)g_1H + \cdots + (a_k + b_k)g_kH \\ &c(a_1g_1H + \cdots +) = (ca_1)g_1H + \cdots + (ca_k)g_kH, \end{aligned}$$

for $a_1, \cdots, a_k, b_1, \cdots, b_k, c \in \mathbb{C}$ and $g_1, \cdots, g_k \in \mathbb{C}$. The action of G on $\mathcal{H}$ can be extend to an action on $\mathbb{C}\mathcal{H}$ by linearity:

$$g(a_1g_1H + \cdots + a_kg_kH) = a_1(gg_1H) + \cdots + a_k(gg_kH), \tag{109}$$

for all $g \in G$. In this way $\mathbb{C}\mathcal{H}$ becomes a G-module of dimension $|\mathcal{H}| = k$. More generally, given a set S on which a group G acts, then the associated module $\mathbb{C}S$ is called the permutation representation associated with S and the elements of S form a basis for $\mathbb{C}S$ called the standard basis. Note that if $H = G$ then the coset representation reduces to the trivial representation. We have the following

Proposition A.1. *Let H be a subgroup of G which has transversal $\{t_1 \cdots t_k\}$ with cosets $\mathcal{H} = t_1H, \cdots t_kH$. Then the matrices of $\mathrm{Ind}_H^G(\mathbf{1})$ are identical with those of G acting on the basis $\mathcal{H}$ for the coset module $\mathbb{C}\mathcal{H}$.*

Proof. Let the matrices for the representations ρ and σ of $\mathrm{Ind}_H^G(\mathbf{1})$ and $\mathbb{C}\mathcal{H}$ be $X = (x_{ij})$ and $Y = (y_{ij})$ respectively. The matrix elements of both matrices are only zeros and ones. Moreover, for any $g \in G$ $x_{ij}(g) = 1$ if and only if $t_i^{-1}gt_j \in H$. But $t_i^{-1}gt_j \in H$ if and only if $gt_jH = t_iH$ and this happens if and only if $z_{ij}(g) = 1$. □

We will now prove the reciprocity law of Frobenius, which relates inner products of restricted and induced characters. Before we need a formula for the character of an induced representation. Let G, H and t_i as in the preceding proposition. Consider a representation ρ of H with character ψ. The transversal t_i give rise to the representation $\mathrm{Ind}_H^G(\rho)$ with character χ. We recall that the character of a representation is $\chi(g) = \mathrm{Tr}\rho(g)$ and given another representation σ with character ψ one can define an inner product of χ and ψ as $\langle\chi,\psi\rangle := \frac{1}{|G|}\sum_{g\in G}\chi(g)\overline{\psi(g)}$, where the bar stands for complex conjugation. Note that $\overline{\psi(g)} = \psi(g^{-1})$. We have $\psi(t_i^{-1}gt_i) = \psi(h^{-1}t_i^{-1}gt_ih)$ for any $h \in H$, so:

$$\chi(\mathrm{Ind}_H^G(\rho(g))) = \sum_i \psi(t_i^{-1}gt_i) = \frac{1}{|H|}\sum_i\sum_{h\in H}\psi(h^{-1}t_i^{-1}gt_ih), \tag{110}$$

but as h runs over H and t_i run over the transversal, the product t_ih runs over all the elements of G exactly once. Thus:

$$\chi(\mathrm{Ind}_H^G(\rho(g))) = \frac{1}{|H|}\sum_{x\in G}\psi(x^{-1}gx). \tag{111}$$

We can now prove the Frobenius reciprocity. To simplify the notation we use $\psi(\mathrm{Ind}_H^G(\rho))$ to indicate the character of the induced representation of ρ if its character is ψ and analogously for the character of the restrict representation we use $\chi(\mathrm{Res}_H^G(\sigma))$ if σ has character χ.

Theorem A.1 (Frobenius Reciprocity). *Let G a group and H a subgroup, ρ a representation of H with character ψ and σ a representation of G with character χ. Then*

$$\langle\psi(\mathrm{Ind}_H^G(\rho)),\chi(\sigma)\rangle = \langle\psi(\rho),\chi(\mathrm{Res}_H^G(\sigma))\rangle, \tag{112}$$

where the left inner product is calculated in G and the right one in H.

Proof. We have the following identities:

$$\begin{aligned}
\langle\psi(\mathrm{Ind}_H^G(\rho)),\chi(\sigma)\rangle &= \frac{1}{|G|}\sum_{g\in G}\psi(\mathrm{Ind}_H^G(\rho))\chi(g^{-1}) = \frac{1}{|G||H|}\sum_{x\in G}\sum_{g\in G}\psi(x^{-1}gx)\chi(g^{-1})\\
&= \frac{1}{|G||H|}\sum_{x\in G}\sum_{y\in G}\psi(y)\chi(xy^{-1}x^{-1}) = \frac{1}{|G||H|}\sum_{x\in G}\sum_{y\in G}\psi(y)\chi(y^{-1})\\
&= \frac{1}{|H|}\sum_{y\in G}\psi(y)\chi(y^{-1}) = \frac{1}{|H|}\sum_{y\in H}\psi(y)\chi(y^{-1})\\
&= \langle\psi(\rho),\chi(\mathrm{Res}_H^G(\sigma))\rangle.
\end{aligned}$$

The second identity follows from equation (111), in the third we posed $y = x^{-1}gx$, the fourth follows from the constancy of χ on the equivalent classes of G, the fifth because x is constant in the sum and the sixth because ψ is zero outside H. □

References

[AMS] J. J. Atick, G. W. Moore, and A. Sen, *Some Global Issues in String Perturbation Theory.*, Nucl. Phys. **B308** (1988) 1–101.

[ARS] J. J. Atick, J. M. Rabin, A. Sen, *An Ambiguity in Fermionic String Perturbation Theory*, Nucl. Phys. **B299** (1988) 279–294.

[BM] A. A. Beilinson and Yu. I. Manin, "The Mumford Form and the Polyakov Measure in String Theory," Commun. Math. Phys. **107** (1986) 359.

[BK] A. A. Belavin and V. G. Knizhnik, "Algebraic Geometry and the Geometry of Quantum Strings," Phys. Lett. B **168** (1986) 201.

[CD] S. L. Cacciatori, F. Dalla Piazza, *Two Loop Superstring Amplitudes and S_6 Representation*, Lett. Math. Phys. **83** (2008) 127–138.

[CDG1] S. L. Cacciatori, F. Dalla Piazza and B. van Geemen, *Modular Forms and Three Loop Superstring Amplitudes*, Nucl. Phys. **B800** (2008) 565–590.

[CDG2] S. L. Cacciatori, F. Dalla Piazza and B. van Geemen, *Genus Four Superstring Measures*, Lett. Math. Phys. **85** (2008) 185–193.

[DP1] E. D'Hoker, D.H. Phong, *Two-Loop Superstrings I, Main Formulas*, Phys. Lett. **B 529** (2002) 241–255.

[DP2] E. D'Hoker, D.H. Phong, *Two Loop Superstrings II. The Chiral measure on Moduli Space*, Nucl. Phys. **B636** (2002) 3–60.

[DP3] E. D'Hoker, D.H. Phong, *Two Loop Superstrings III. Slice Independence and Absence of Ambiguities*, Nucl. Phys. **B636** (2002) 61–79.

[DP4] E. D'Hoker, D.H. Phong, *Two-Loop Superstrings IV: The Cosmological Constant and Modular Forms*, Nucl. Phys. **B 639** (2002) 129–181.

[DP5] E. D'Hoker, D.H. Phong, *Asyzygies, Modular Forms, and the Superstring Measure I*, Nucl. Phys. B **710** (2005) 58–82.

[DP6] E. D'Hoker, D.H. Phong, *Asyzygies, Modular Forms, and the Superstring Measure II*, Nucl. Phys. B **710** (2005) 83–116.

[DP7] E. D'Hoker and D. H. Phong, *The Geometry of String Perturbation Theory*, Rev. Mod. Phys. **60** (1988) 917.

[DP8] E. D'Hoker and D. H. Phong, *Superstrings, Super Riemann Surfaces, and Supermoduli Space*, Contribution to the Proc. of thr 'String Theory' Conf., Rome, Italy, Jun 1988. Published in Rome String Theory 1988:17-68 (QCD162:S75:1988)

[DvG] F. Dalla Piazza, Bert van Geemen, *Siegel Modular Forms and Finite Symplectic Groups*, arXiv:0804.3769.

[DBMS] P. Dunin-Barkowski, A. Morozov and A. Sleptsov, "Lattice Theta Constants vs Riemann Theta Constants and NSR Superstring Measures," arXiv:0908.2113 [hep-th].

[F] J. Fay *Theta Functions on Riemann Surfaces*, Lect.Notes Math. 352, Springer, 1973

[F1] J.S. Frame, *The Classes and Representations of the Group of 27 Lines and 28 Bitangents*, Annali di Mathematica Pura ed Applicata, **32** (1951) 83–119.

[F2] J.S. Frame, *Some Characters of Orthogonal Groups Over the Field of Two Elements* , In: Proc. of the Second Inter. Conf. on the Theory of Groups, Lecture Notes in Math., Vol. 372, pp. 298–314, Springer, 1974.

[GN1] S. B. Giddings and P. C. Nelson, "The Geometry of Superriemann Ssurfaces," Commun. Math. Phys. **116** (1988) 607.

[GN2] S. B. Giddings and P. C. Nelson, "Line Bundles on Superriemann Surfaces," Commun. Math. Phys. **118** (1988) 289.

[Gr] S. Grushevsky, *Superstring Scattering Amplitudes in Higher Genus* , Commun. Math. Phys. 287 (2009) 749–767.

[GSM] S. Grushevsky and R. Salvati Manni, "The Vanishing of Two-Point Functions for Three-Loop Superstring Scattering Amplitudes," arXiv:0806.0354 [hep-th].

[J] N. Jacobson, Basic Algebra I. W.H. Freeman and Company, 1974.

[I1] J. Igusa, Theta functions, Springer 1972.

[I2] J. Igusa, *On the graded ring of theta-constants* , Amer. J. Math. **86** (1964) 219–246.

[I3] J. Igusa, *Schottky's invariant and quadratic forms* , in: International Christoffel Symposium proceedings, editors P.L. Butzer, F. Fehér, Birkhäuser, 1981:352–362.

[L] S. Lang, "Algebra," Graduate Text in Mathematics, Springer-Verlag (2002 New York).

[MYu1] Yu. I. Manin, "The Partition Function of the Polyakov String can be expressed in Terms of Theta Functions," Phys. Lett. B **172** (1986) 184.

[MYu2] Yu. I. Manin, "Theta Function Representation of the Partition Function of a Polyakov String," JETP Lett. **43** (1986) 204 [Pisma Zh. Eksp. Teor. Fiz. **43** (1986) 161].

[Ma1] E. J. Martinec, “Nonrenormalization Theorems and Fermionic String Finiteness,” Phys. Lett. B **171** (1986) 189.

[Ma2] E. J. Martinec, “Conformal Field Theory on a (Super)Riemann Surface,” Nucl. Phys. B **281** (1987) 157.

[MV] M. Matone and R. Volpato, “Superstring Measure and Non-Renormalization of the Three-Point Amplitude,” Nucl. Phys. B **806** (2009) 735

[MM] G. W. Moore and A. Morozov, *Some Remarks on Two Loop Superstring Calculations*, Nucl. Phys. B **306** (1988) 387.

[Mo] A. Morozov, “NSR Superstring Measures Revisited,” JHEP **0805** (2008) 086

[Mo1] A. Morozov, “NSR Measures on Hyperelliptic Locus and Non-Renormalization of 1,2,3-Point Functions,” Phys. Lett. B **664** (2008) 116

[Mo2] A. Morozov, “Pointwise Vanishing of Two Loop Contributions to 1, 2, 3 Point Functions in the NSR Formalism,” Nucl. Phys. B **318** (1989) 137 [Theor. Math. Phys. **81** (1990 TMFZA,81,24-35.1989) 1027].

[MP] A. Morozov and A. Perelomov, “A Note on Multiloop Calculations for Superstrings in the NSR Formalism,” Int. J. Mod. Phys. A **4** (1989) 1773 [Sov. Phys. JETP **68** (1989 ZETFA,95,1153-1161.1989) 665.1989 ZETFA,95,1153].

[Mu] D. Mumford, *Tata lectures on theta. II.*

[Mu2] D. Mumford, “Stability of Projective Varieties,” Enseign. Math. **23**, (1977) 39-100.

[OSM] M. Oura, R. Salvati Manni, *On the Image of Code Polynomials Under Theta Map*, arXiv:0803.4389.

[OPSMY] M. Oura, C. Poor, R. Salvati Manni and D. Yuen, “Modular Forms of Weight 8 for $\Gamma_g(1,2)$, arXiv:0811.2259

[RF] H. E. Rauch and H. M. Farkas, Theta Functions with Applications to Riemann Surfaces. The Williams & Wilkins Company, 1974.

[R1] B. Runge, *On Siegel Modular Forms, Part I*, J. Reine Angew. Math. **436** (1993) 57–85.

[R2] B. Runge, *On Siegel Modular Forms, Part II*, Nagoya Math. J. **138** (1995) 179–197.

[S] B. E. Sagan, The Symmetric Group. Representations, Combinatorial Algorithms, and Symmetric Funtions. Springer 2001.

[SM1] R. Salvati Manni, *Modular Varieties with Level* 2 *Theta Structure*, Amer. J. Math. **116** (1994) 1489–1511.

[SM2] R. Salvati Manni, *Remarks on Superstring Amplitudes in Higher Genus* , Nucl. Phys. **B801** (2008) 163–173.

[vG1] B. van Geemen, *The Schottky problem and second order theta functions* , Workshop on Abelian Varieties and Theta Functions (Spanish) (Morelia, 1996), 41–84, Aportaciones Mat. Investig., 13, Soc. Mat. Mexicana, Mexico, 1998.

[vG2] B. van Geemen, *Siegel Modular Forms Vanishing on the Moduli Space of Curves* , Invent. Math. **78** (1984) 329–349.

[vGvdG] B. van Geemen, G. van der Geer, *Kummer Varieties and Moduli Spaces of Abelian Varieties* Am. J. Math. **108** (1986) 615–642.

[VV] E. P. Verlinde and H. L. Verlinde, "Multiloop Calculations in Covariant Superstring Theory," Phys. Lett. B **192** (1987) 95.

In: Superstring Theory in the 21st Century...
Editor: Gerold B. Charney, pp.87-124
ISBN 978-1-61668-385-6

Chapter 3

THE COSMOLOGY OF THE TYPE IIB SUPERSTRINGS THEORY WITH FLUXES

Matos Tonatiuh[a*], ***José-Rubén Luévano***[b], ***Hugo García-Compeán***[a] ***and Erandy Ramírez***[c]
[a]Departamento de Física, Centro de Investigación y de Estudios Avanzados del IPN, A.P. 14-740, 07000 México D.F., México.
[b]Departamento de Ciencias Básicas, Universidad Autónoma Metropolitana-Azcapotzalco, C.P. 02200 México, D.F., México.
[c]Fakultät für Physik, Universität Bielefeld, Universitätstrasse 25, Bielefeld 33615, Germany

Abstract

In the tipical superstrings cosmology, the dilaton is usually interpreted as a Quintessence field. This work is a review of an alternative interpretation of the dilaton, namely, as the dark matter of the universe, in the context of a particular cosmological model derived from type IIB supergravity theory with fluxes. First we study the conditions needed to have an early epoch of inflationary expansion with a potential coming from type IIB superstring theory with fluxes involving two moduli fields; the dilaton and the axion. The phenomenology of this potential is different from the usual hybrid inflation scenario and we analyze the possibility that the system of field equations undergo a period of inflation in three different regimes with the dynamics modified by a Randall-Sundrum term in the Friedmann equation. We find that the system, can produce inflation and due to the modification of the dynamics, a period of accelerated contraction can follow or preceed this inflationary stage depending on the sign of one of the parameters of the potential. We discuss on the viability of this model in a cosmological context. With this alternative interpretation we also find that the model gives a similar evolution and structure formation of the universe compared with the ΛCDM model in the linear regime of fluctuations of the structure formation. Some free parameters of the theory are fixed using the present cosmological observations. In the non-linear regime there are some differences between the type IIB supergravity theory with the traditional CDM paradigm. The supergravity theory predicts the formation of galaxies earlier than the CDM and there is no density cusp in the centre of galaxies. These differences can distinguish both models and might give a distinctive

[*]Part of the Instituto Avanzado de Cosmología (IAC) collaboration http://www.iac.edu.mx/

feature to the phenomenology of the cosmology coming from superstring theory with fluxes.

PACS numbers: 11.25.Wx,95.35.+d,98.80.-k

1. Introduction

One of the main problems of superstring theory is that there is not a real phenomenology which can support this theory. Usually, superstring theory is supported only by its mathematical and internal consistency, but not by real experiments or observations. For some people, like the authors, one way of how superstring theory can make contact with phenomenology is through the cosmology [3]. In the last years, a number of new observations have given rise to a new cosmology and to a new perception of the universe (see for example [4]). In superstring theory there are 6 extra dimensions forming a compact internal Calabi-Yau manifold [5]. Size and shape of this manifold manifests, at the four dimensional low energy effective field theory, a series of scalar fields (moduli of the theory) many of which apparently have not been seen in nature. In particular, two fields, the *dilaton* and the *axion*, are two very important components of the theory which can not be easily fixed. In fact, one should find a physical interpretation for these fields or give an explanation of why we are not able to see them in nature.

One of the main problems in physics now is to know the nature of the dark matter and the understanding of the accelerated expansion of the universe. These two phenomena have been observed in the last years and now there are a number of observations supporting the existence of the dark matter [1] and the accelerated expansion of the universe as well [2].

The last 50 years have been some of the most fruitful ones in the life of physics, the standard model of particles (SM) and the standard model of cosmology (SMC) essentially developed in this period, are now able to explain a great number of observations in laboratories and cosmological observatories as never before. In only 50 years there have been great steps in the understanding of the origin and development of the universe. Nevertheless, many questions are still open, for example the SMC contains two periods, inflation and structure formation. For the understanding of the structure formation epoch we need to postulate the existence of two kind of substances, the dark matter and the dark energy. Without them it is impossible to explain the formation of galaxies and clusters of galaxies, or the observed accelerated expansion of the universe. On the other hand, it has been postulated a period of inflation in order to give an explanation for several observations as the homogeneity of the universe, the close value of the density of the universe to the critical density or the formation of the seeds which formed the galaxies. However, there is not a theory that unifies these two periods, essentially they are disconnected from each other.

One interpretation of the moduli fields in superstrings theory is that there exist a mechanism for eliminating these fields during the evolution of the universe [6]. Recently, one of the most popular interpretations for the dilaton field is that it can be the dark energy of the universe, *i.e.* a Quintessence field [7]. These last interpretations have been possible because after a non trivial compactification, the dilaton field acquires an effective potential. This effective potential makes possible to compare the dilaton field with some other kinds of matter [7]. In this work we are giving the dilaton a *different* interpretation supposing that it is the dark matter instead of the dark energy [8]. Such attempts have been carried over in the past with other dilaton potentials [9]. Here, we will be very specific starting with an effective potential derived recently from the type IIB supergravity theory. The main goal of this work is to show that this interpretation could be closer to a realistic cosmology as the interpretation that the dilaton is the dark energy. We will see that the late cosmology is very similar to the ΛCDM one with this alternative interpretation. Nevertheless, we will also see that it is necessary to do something else in order to recover a realistic cosmology from superstring theory. On the other hand, a great deal of work has been done recently, in the context of string compactifications with three-form fluxes (R-R and NS-NS) on the internal six-dimensional space and the exploration on their consequences in the stabilization of the moduli fields including the dilaton Φ and axion C [10]. Moduli stabilization has been used also in string cosmology to fix other moduli fields than the volume modulus including dilaton+axion and Kahler moduli [11]. For a description of more realistic scenarios, see [12].

In the context of the type IIB supergravity theory on the $\mathbf{T}^6/\mathbb{Z}_2$ orientifold with a self-dual three-form fluxes, it has been shown that after compactifaying the effective dilaton-axion potential is given by [29]

$$
\begin{aligned}
V_{dil} &= \frac{M_P^4}{4(8\pi)^3} h^2 \, e^{-2\Sigma_i \sigma_i} \left[e^{-\Phi^{(0)}} \cosh\left(\Phi - \Phi^{(0)}\right) \right. \\
&+ \left. \frac{1}{2} e^{\Phi} (C - C^{(0)})^2 - e^{-\Phi^{(0)}} \right], \qquad (1)
\end{aligned}
$$

where $h^2 = \frac{1}{6} h_{mnp} h_{qrs} \delta^{mq} \delta^{nr} \delta^{ps}$. Here h_{mnp} are the NS-NS integral fluxes, the superscript (0) in the fields stands for the fields in the vacuum configuration and finally σ_i with $i = 1, 2, 3$ are the overall size of each factor $\mathbf{T}^2$ of the $\mathbf{T}^6/\mathbb{Z}_2$ orientifold (in [29] there is a mistprint in the potential 1). Here we will simplify the system supposing that the moduli fields σ_i are constant for the late universe.

For the sake of simplicity in the derivation of the potential (1), some assumptions were made [29]. One of them is the assumption that the tensions of D-branes and orientifold planes cancel with the energy V_{dil} at $\Phi = \Phi_0$ and $C = C_0$. An assumption on initial conditions is that the dilaton is taken to deviate from equilibrium value, while the complex structure moduli are not. It is also assumed that (1) has a global minimum Φ_0, such that $V(\Phi_0) = 0$. Also that, the complex moduli are fixed and only the radial modulus σ feels a potential when the dilaton-axion system is excited. These assumptions make the model more simple, but still with the sufficient structure to be of interest in cosmological and astrophysical problems.

In order to study the cosmology of this model, it is convenient to define the following quantities $\lambda\sqrt{\kappa}\phi = \Phi - \Phi^{(0)}$, $V_0 = \frac{M_P^4}{4(8\pi)^3} h^2 \, e^{-2\Sigma_i \sigma_i} \, e^{-\Phi^{(0)}}$, $C - C^{(0)} = \sqrt{\kappa}\psi$ and $\psi_0 = e^{\Phi^{(0)}}$, where λ is the string coupling $\lambda = e^{\langle \Phi \rangle}$ and $\lambda\sqrt{2\kappa}$ is the reduced Plank mass $M_p/\sqrt{8\pi}$. With this new variables, the dilaton potential transforms into

$$
\begin{aligned}
V_{dil} &= V_0 \left(\cosh\left(\lambda\sqrt{\kappa}\phi\right) - 1\right) + \frac{1}{2} V_0 \, e^{\lambda\sqrt{\kappa}\phi} {\psi_0}^2 \kappa \psi^2 \\
&= V_\phi + e^{\lambda\sqrt{\kappa}\phi} V_\psi. \qquad (2)
\end{aligned}
$$

In some works the scalar field potential (2) is suggested to be the dark energy of the universe, that means, a Quintessence field [7][29]. In this work we are not following this interpretation to the dilaton field. Instead of this, we will interpret the term $V_0\left(\cosh\left(\lambda\sqrt{\kappa}\phi\right)-1\right)$ as the dark matter of the universe[14], [15]. The remaining term in V_{dil} contains the contribution of the axion field C. This is what makes the difference between our work and previous ones. This interpretation allows us to compare the cosmology derived from the potential (2) with the Λ cold dark matter (ΛCDM) model. The rest of the fields coming from superstrings theory can be modeled as usual, assuming that this part of the matter is a perfect fluid. This perfect fluid has two epochs: radiation and matter dominated ones. In order to consider both epochs we write the matter component as matter and radiation, with a state equation given by $\dot{\rho}_b + 3\,H\,\rho_b = 0$ and $\dot{\rho}_{rad} + 4\,H\,\rho_{rad} = 0$. For modelling the dark energy we can take the most general form supposing that it is also a perfect fluid with the equation of state given by $\dot{\rho}_L + 3\,\gamma_{DE}\,H\,\rho_L = 0$, where γ_{DE} is smaller than 1/3 and can even be negative in the case it represents a phantom energy [16] field. It is just zero if ρ_L represents the cosmological constant $L = \rho_L$.

2. Inflation

In this section we study the possibility that superstring theory could account for the unification between inflation and the structure formation using a specific example. Recently Frey and Mazumdar [29] were able to compactify the IIB superstrings including 32 fluxes. Earlier works discussing string flux compactifications are [30][31][32][33] [34][35][36], and works addressing the issue that the inflaton carries the SM gauge charges as opposed to an arbitrary gauge singlet in the context of string theory can be found in [37][38]. In the context of the type IIB supergravity theory on the $\mathbf{T}^6/\mathbb{Z}_2$ orientifold with a self-dual three-form fluxes, it has been shown that after compactifaying, the effective dilaton-axion potential is given by (2),*

This potential contains two main scalar fields (moduli fields), the dilaton Φ and the axion C. In [39] the dilaton was interpreted as dark matter and under certain conditions the model reproduces the observed universe, $i.e.$ the structure formation. In this work we investigate if the same theory could give an inflationary period in order to obtain a unified picture between these two epochs. In other words, in this section we search if it is possible that the same low energy Lagrangian of IIB superstrings with the scalar and axion potential (1) can give an acceptable inflationary period. To handle the fluxes, we work in the brane representation of space-time, working with a RS-II modification in the equations. Thus, due to the presence of the fluxes during this period, these models have the phenomenology of the Randall-Sundum models [44],[45],[29] and we have chosen to work with the RS-II one, but the same type of analysis could be done for the other model. An important point to mention before discussing the equations is that we are holding the σ_i fixed, not considering at the moment a potential to stabilize them. In particular, the modification of the dynamics into a Randall-Sundrum one is a consequence of this approximation †. The paper is organized as

*We thank Andrew Frey for pointing out a mistake in the previous version of the potential. This was due to a typographical error in [29] and it does not affect our work.

†Thanks to Andrew Frey for clarifying this point.

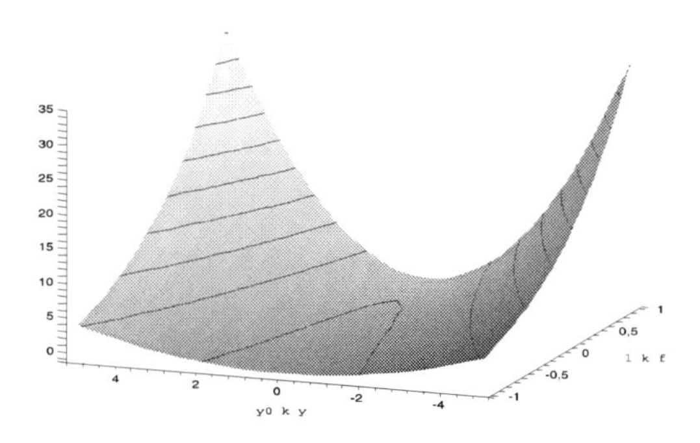

Figure 1. Potential

follows: in section 3. we introduce an appropriate parametrization of the potential 1 in order to give unities to physical quantities. As a first insight to study the cosmology of the low energy Lagrangian, we make an approximation to the potential and give the field equations leaving the analysis of the complete system to section 5.. In section 4. we find the solutions of the equations in different regimes and conditions and the results they give. In section 5., the dynamical evolution of the full system is performed for large values of the field and in section 7. we discuss some conclusions and perspectives.

3. The Potential

In what follows we want to study the behavior of the potential (2) at early times when the scalar field takes large values and study the conditions that the parameters λ and V_0 need to meet in order to have inflation. Thus expressing the cosh function in terms of exponentials and taking the limit ϕ big, we arrive to the following expression for the potential:

$$V(\phi,\psi) = \frac{1}{2} V_0 e^{\lambda\sqrt{\kappa}\phi}(1 + \kappa\psi_0^2\psi^2) - V_0, \tag{3}$$

From the approximation, there remains a term V_0 which acts during the inflationary period like a "cosmological constant" and that we will address as a free parameter of the model. In contrast with the usual hybrid inflation scenario [40], there is no critical value for which this potential exhibits a phase transition triggering the end of inflation (if any such process occurs). The potential follows an exponential behavior in the ϕ field that prevents it from staying at a fixed value from the start, i.e. it cannot relax at $\phi = 0$ or at any other different value (apart from infinity). A plot of the potential illustrates this behavior, see Fig. 1. Then we will assume in this section and the next, that there is some mechanism by which the ψ field rolls down to its minimum at $\psi = 0$ oscillating around it at the very early stages of evolution and that any processes such as inflation took place afterwards. So any information concerning such a process is erased by the expansion led by the ϕ field if inflation is to happen. Thus we can work with the following expression for the potential

$$V(\phi) = \frac{1}{2} V_0 e^{\lambda\sqrt{\kappa}\phi} - V_0 \tag{4}$$

We will follow closely the analysis done by Copeland *et al.* [41] and Mendes and Liddle [42] in order to obtain the conditions for this potential to undergo inflation in the cases when the scalar field or the V_0 term dominate the dynamics as well as in the intermediate stage.

Our calculations are performed in the high-energy regime within the slow-roll approximation since a potential slow-roll formalism has already been provided for this scenario [47].

3.1. Field equations

In the presence of branes, in the RS-II scenario, the Friedmann equation for this cosmology changes from its usual expression to [45]:

$$H^2 = \frac{\kappa}{3}\rho\left(1+\frac{\rho}{\rho_0}\right), \tag{5}$$

where $H \equiv \dot{a}/a$, a is the scale factor of the Universe, a dot means derivative with respect to time and ρ_0 is the brane tension. The total density ρ as well as the equations of motion for the fields for the standard cosmology case are deduced in [46]:

$$\rho = \frac{1}{2}\dot{\phi}^2 + \frac{1}{2}\dot{\psi}^2 e^{\lambda\sqrt{\kappa}\phi} + V_\phi + V_\psi e^{\lambda\sqrt{\kappa}\phi}. \tag{6}$$

In our case,

$$V_\phi = \frac{1}{2}V_0 e^{\lambda\sqrt{\kappa}\phi} - V_0, \quad V_\psi = \frac{1}{2}V_0\kappa\psi_0^2\psi^2 \tag{7}$$

which in the slow-roll approximation can be written as

$$H^2 \simeq \frac{\kappa}{3}\left(V_\phi + V_\psi e^{\lambda\sqrt{\kappa}\phi}\right)\left[1+\frac{\left(V_\phi + V_\psi e^{\lambda\sqrt{\kappa}\phi}\right)}{\rho_0}\right] \tag{8}$$

The equations of motion for both fields are given considering only the presence of both scalar fields with no radiation fluid, they are [46]

$$\begin{aligned} \ddot{\phi} + 3H\dot{\phi} + \frac{\partial V_\phi}{\partial \phi} &= \lambda\sqrt{\kappa}e^{\lambda\sqrt{\kappa}\phi}\left(\frac{1}{2}\dot{\psi}^2 - V_\psi\right) \\ \ddot{\psi} + 3H\dot{\psi} + \frac{\partial V_\psi}{\partial \psi} &= -\lambda\sqrt{\kappa}\dot{\phi}\dot{\psi}. \end{aligned} \tag{9}$$

Which in the slow-roll approximation read:

$$\begin{aligned} 3H\dot{\phi} + \frac{\partial V_\phi}{\partial \phi} &\simeq -\lambda\sqrt{\kappa}e^{\lambda\sqrt{\kappa}\phi}V_\psi \\ 3H\dot{\psi} + \frac{\partial V_\psi}{\partial \psi} &\simeq 0. \end{aligned} \tag{10}$$

since the right-hand side of the second equation in (9) can be taken as a kinetic term to the square.

A consequence of considering the ψ field in the value corresponding to the minimum of its potential ($\psi = 0$) is that the system of equations does not have source terms and couplings of the fields. This allows to do an inflationary analysis equivalent to the usual one in terms of the potential slow-roll parameters for the RS-II model. But this restriction lacks the information about how the interaction of both fields affects the dynamics. The results presented in these sections are therefore a particular case of a more general analysis presented in section 5..

4. Inflationary Phenomenology of the Model

The analysis that follows will be confined to the high-energy limit of this model, which simplifies calculations further. Consequently, we have the following system of equations within the slow-roll approximation and the high-energy limit:

$$H \simeq \sqrt{\frac{\kappa}{3\rho_0}} V;$$
$$3H\dot{\phi} + \frac{\partial V_\phi}{\partial \phi} \simeq 0 \tag{11}$$

which can be solved analytically. In this work we take the convention $\dot{\phi} < 0$, hence the ϕ field is a decreasing function of time. The solutions of the field equations will be given in section 4.6..

The expressions for the potential slow-roll parameters having the inflation field confined to the brane in the intermediate regime and high-energy limit of this cosmology were deduced by Maartens *et al.* [47]. For the high-energy limit they are :

$$\epsilon \simeq \frac{1}{\kappa}\left(\frac{V'}{V}\right)^2 \frac{\rho_0}{V}; \qquad \eta \simeq \frac{1}{\kappa}\left(\frac{V''}{V}\right)\frac{\rho_0}{V} \tag{12}$$

where primes indicate derivatives with respect to the ϕ field. The slow-roll approximation is satisfied as long as the slow-roll parameters defined previously accomplish the following conditions :

$$\epsilon \ll 1, \qquad |\eta| \ll 1. \tag{13}$$

The number of e-foldings of inflation in terms of the potential for this model is given in our notation by [47]:

$$N \simeq -\frac{\kappa}{\rho_0}\int_{\phi_N}^{\phi_e} \frac{V^2}{V'} d\phi, \tag{14}$$

where ϕ_N represents the value of the field ϕ at N e-foldings of expansion before the end of inflation and ϕ_e is the value at the end of inflation.

For the potential (4) we have

$$\epsilon = \frac{\rho_0 \lambda^2}{4V_0} \frac{e^{2\lambda\sqrt{\kappa}\phi}}{\left(\frac{e^{\lambda\sqrt{\kappa}\phi}}{2} - 1\right)^3} \tag{15}$$

$$\eta = \frac{\rho_0 \lambda^2}{2V_0} \frac{e^{\lambda\sqrt{\kappa}\phi}}{\left(\frac{e^{\lambda\sqrt{\kappa}\phi}}{2} - 1\right)^2} \tag{16}$$

$$\begin{aligned} N &\simeq \frac{2V_0}{\lambda^2 \rho_0} \left[-\frac{1}{4} \left(e^{\lambda\sqrt{\kappa}\phi_e} - e^{\lambda\sqrt{\kappa}\phi_N} \right) \right. \\ &+ \left. \left(e^{-\lambda\sqrt{\kappa}\phi_e} - e^{-\lambda\sqrt{\kappa}\phi_N} \right) + \lambda\sqrt{\kappa}(\phi_e - \phi_N) \right]. \end{aligned} \tag{17}$$

Since, in our model there is no value of ϕ that may lead ψ to a global minimum, the only way in which inflation can finish is by the violation of the slow-roll approximation with ϵ exceeding unity.

The value of ϕ at which ϵ becomes equal to unity is

$$\begin{aligned} \sqrt{\kappa}\phi_e &= \frac{1}{\lambda} \ln \left\{ \left[\frac{2\lambda^2 \rho_0}{3V_0} \right] \left[\frac{4^{1/3} B^{1/3}}{2\lambda^2 \rho_0} + \frac{(6V_0 + \lambda^2 \rho_0) 4^{2/3}}{2B^{1/3}} \right. \right. \\ &+ \left. \left. \frac{(3V_0 + \lambda^2 \rho_0)}{\lambda^2 \rho_0} \right] \right\} \end{aligned} \tag{18}$$

with

$$\begin{aligned} B &= \lambda^2 \rho_0 \left[27V_0^2 + 18V_0 \lambda^2 \rho_0 + 2\lambda^4 \rho_0^2 \right. \\ &+ \left. (3V_0)^{3/2} \sqrt{4\lambda^2 \rho_0 + 27V_0} \right]. \end{aligned} \tag{19}$$

We can rearrange the previous expression so that

$$\begin{aligned} \sqrt{\kappa}\phi_e &= \frac{1}{\lambda} \ln \left\{ \left[\frac{2\lambda^2 \rho_0}{3V_0} \right] \left[1 + \frac{3V_0}{\lambda^2 \rho_0} \right. \right. \\ &+ \left. \left. \frac{2^{2/3} B^{1/3}}{2\lambda^2 \rho_0} + \frac{(6V_0 + \lambda^2 \rho_0) 2^{1/3}}{B^{1/3}} \right] \right\} \end{aligned} \tag{20}$$

So we have that if

$$\frac{3V_0}{\lambda^2 \rho_0} + \frac{B^{1/3}}{2^{1/3} \lambda^2 \rho_0} + \frac{2^{1/3}(6V_0 + \lambda^2 \rho_0)}{B^{1/3}} \ll 1, \tag{21}$$

the ϕ field dominates the potential Eq (4) and we will have an exponential potential, which in the standard cosmology case corresponds to power-law inflation, but not for the RS-II modification.

The bound found before, depends on the choice of values for λ and V_0. As we will see in section 4.2., λ is fixed once the value of the brane tension is given. So actually Eq (21) depends only on the choice of V_0. We use the computing package Mathematica to find a value of V_0 that satisfies the condition (17). This happens if:

$$\lambda^2 \rho_0 < 0, \quad \Longrightarrow \quad -\frac{4\lambda^2 \rho_0}{27} \leq V_0 < -\frac{\rho_0 \lambda^2}{6} \tag{22}$$

$$\lambda^2 \rho_0 > 0, \quad \Longrightarrow \quad V_0 < -\frac{1}{6} \rho_0 \lambda^2 \tag{23}$$

From these the second case satisfies having a brane tension with no incompatibilities with nucleosynthesis [42],[50].

4.1. Density Perturbations

The field responsible for inflation produces perturbations which can be of three types: scalar, vector and tensor. Vector perturbations decay in an expanding universe and tensor perturbations do not lead to gravitational instabilities that may lead to structure formation. The adiabatic scalar or density perturbations can produce these type of instabilities through the vacuum fluctuations of the field driving the inflationary expansion. So they are usually thought to be the seed of the large scale structures of the universe. One of the quantities that determined the spectrum of the density perturbations is δ_H which gives the density contrast at horizon crossing (if evaluated at that scale). For the RS-II modification and in our notation, this quantity is [42],[48]

$$\delta(k)_H^2 \simeq \frac{\kappa^3}{75\pi^2}\frac{V^3}{V'^2}\frac{V^3}{\rho_0^3}. \tag{24}$$

Evaluated at the moment of horizon crossing when the scale $k = aH$. The slow-roll approximation guarantees that δ_H is nearly independent of scale when scales of cosmological interest are crossing the horizon. The reason being that the field is almost constant in time. satisfying the new COBE constrain updated to $\delta_H = 1.9 \times 10^{-5}$ [51]. Here we take 60 e-foldings before the end of inflation to find the scales of cosmological interest.

We use equation (17),provided we know the value of ϕ_e given by equation (18), to find the value ϕ_N corresponding to $N = 60$, that is 60 e-foldings before the end of inflation and evaluate δ_H as

$$\delta_H^2 \simeq \frac{4\kappa^2}{75\pi^2}\frac{V_0^4}{\lambda^2\rho_0^3}e^{-2\lambda\sqrt{\kappa}\phi_{60}}\left(\frac{e^{\lambda\sqrt{\kappa}\phi_{60}}}{2} - 1\right)^6 \tag{25}$$

The results are given in table 2 of section 4.7.

4.2. Field-Dominated Region

Considering the case when the ψ field plays no role and ϕ governs the dynamics of the expansion alone, from Eq (4), we have a potential of exponential type resembling that of power-law inflation [49]. The first term of Eq (4) dominates giving an exponential expansion but not to power-law inflation because the dynamics of RS-II changes this condition. The slow-roll parameters are given by:

$$\epsilon = \eta \simeq \frac{2\rho_0}{V_0}\frac{\lambda^2}{e^{\lambda\sqrt{\kappa}\phi}}. \tag{26}$$

In contrast with the standard cosmology where they are not only the same but constant. The fact that the dynamics is modified due to the Randall-Sundrum cosmology allows the existence of a value of ϕ that finishes inflation, since as we just saw, the potential slow-roll parameters are equal but not constant. Showing a dependence on ϕ and therefore, an evolution.

We have that in this regime, the value of ϕ_e corresponding to the end of inflation is given by $\epsilon \simeq 1$. The $\simeq$ is used because we are in the potential slow-roll approximation not the Hubble one [43]

$$\sqrt{\kappa}\phi_e \simeq \frac{1}{\lambda}\ln\left(\frac{2\lambda^2\rho_0}{V_0}\right) \tag{27}$$

Inserting this value in the expression for the number of e-foldings (14) for this regime and evaluating, we obtain that ϕ_N with $N = 60$ is

$$\sqrt{\kappa}\phi_{60} \simeq \frac{1}{\lambda} \ln\left(122\frac{\lambda^2 \rho_0}{V_0}\right), \tag{28}$$

and we can evaluate the density contrast Eq (24) for this regime:

$$\delta_H^2 \simeq \frac{61^4 \kappa^2 \lambda^6 \rho_0}{75\pi^2}. \tag{29}$$

One can observe from this equation that given the value of δ_H from observations, it is possible to completely constrain λ as

$$\lambda^6 \simeq \frac{75\pi^2 \delta_H^2}{61^4 \kappa^2 \rho_0} \tag{30}$$

We have a dimensionless number that fixes one of the parameters of the potential and can be contrasted with the value predicted by this supergravity model when interpreted as Dark Matter [39].

If we substitute the last equation into Eq (21), we get in principle a set of values for the constant V_0 that satisfy the field domination condition.

4.3. Vacuum Energy-Dominated Regime

We consider now the regime in which the second term of Eq (4) dominates the dynamics. In this case the slow -roll parameters Eqns (12) are:

$$\epsilon \simeq -\frac{\rho_0 \lambda^2 e^{2\lambda\sqrt{\kappa}\phi}}{4V_0} \tag{31}$$

$$\eta \simeq \frac{\rho_0 \lambda^2 e^{\lambda\sqrt{\kappa}\phi}}{2V_0} \tag{32}$$

We find ourselves here with the fact that ϵ is negative, that is, with a period of deflation [52]

It is necessary to point out that this is a consequence of the modification of the dynamics. The definition of ϵ for the Randall-Sundrum II cosmology in the high-energy limit is not positive definite as in the standard cosmology case. Thus a stage of accelerated contraction for this regime on the potential is only a result of the modification in the field equations.

Following the evolution of the dynamics with the potential Eq (4), from a region where ϕ dominates, to a stage in which the energy V_0 drives the behavior of the expansion, we observe a primordial inflationary expansion that erases all information concerning any process that the ψ field is undergoing under the influence of the potential Eq (3). The intermediate regime, in which both terms in Eq (4) are of the same order, produces further expansion. Finally the field ϕ reaches a value on the potential that commences a stage of accelerated contraction. This value is obtained when the denominator in Eq (15) changes sign:

$$\sqrt{\kappa}\phi_d = \frac{\ln 2}{\lambda}, \tag{33}$$

corresponding to the value of the vacuum-dominated regime. Such process takes place when the field ϕ takes on values below $\sqrt{\kappa}\phi_d$. Substituting the value of ϕ_d in the potential (4), we have that $V(\phi_d) = 0$. Thus the balance of the terms in Eq (4) and the sign of V_0 determine the place where deflation starts as the point where the potential crosses the ϕ axis.

In principle, there would not be a physical reason that could prevent this deflationary stage to stop. But the argument mentioned before, concerning the modification of the dynamics applies again . We observe ϵ has a dependence on the field and therefore undergoes an evolution accordingly. The condition to end deflation is $\epsilon = -1$, in opposition to inflation. In consequence, we can also find from the first of equations (31) a value of ϕ corresponding to this;

$$\epsilon = -1 \Rightarrow \quad \sqrt{\kappa}\phi_e = \frac{1}{2\lambda} \ln\left(\frac{4V_0}{\rho_0\lambda^2}\right). \tag{34}$$

Where this time, the subscript "e", indicates the end of deflation. This value, as we can see, depends on V_0 and λ.

The choice of V_0 parameter will be given in section 4.7. where we give different values according to the conditions we find in the following. We will see whether or not an inflationary stage takes place under the value of λ found in the previous analysis.

4.4. Intermediate Regime

The intermediate regime corresponds to the region where both terms in Eq (4) are of the same order. In order to obtain a bound on the values of V_0 that satisfy the COBE constrain (25), we need to solve numerically Eqs (17) and (18).Since we have that both terms in Eq (4) are important, this means that the exponential is of $\mathcal{O}(1)$, thus we can expand it in Taylor series as $\exp(\lambda\sqrt{\kappa}\phi) \simeq 1 + \lambda\sqrt{\kappa}\phi$ and arrive to a value of ϕ_f equal to

$$\phi_e = \frac{(4B)^{1/3}}{a\lambda V_0} + \left(2 + \frac{\lambda^2\rho_0}{3V_0}\right)\frac{\lambda\rho_0 4^{2/3}}{\kappa B^{1/3}} + \frac{2\lambda\rho_0}{3\sqrt{\kappa}V_0} + \frac{1}{\lambda\sqrt{\kappa}} \tag{35}$$

where B now is given by

$$\begin{aligned} B &= \frac{\rho_0\lambda^2}{\kappa^3}\left(18\rho_0\lambda^2 V_0 + 27V_0^2\right. \\ &+ \left. 2\rho_0^2\lambda^4 + \sqrt{(3V_0)^3(4\rho_0\lambda^2 + 27V_0)}\right) \end{aligned} \tag{36}$$

In order to have a real scalar field, we find two bounds for the values that V_0 can take on following from the roots in the previous expressions:

$$V_0 < -\frac{1}{6}\rho_0\lambda^2, \qquad V_0 > -\frac{4}{27}\rho_0\lambda^2. \tag{37}$$

The first bound coincides with the value given by Eq (23) needed to have field domination. It is an upper bound for the allowed values of V_0 in expression (18). On the other hand, we have from Eqs. (18) and (19) that positive values of V_0 also satisfy that there exists a real value of ϕ_e in the intermediate regime. But there is another condition coming from Eq (19) in which we see that V_0 cannot be 0. So we have an interval for allowed values of V_0 as:

$$V_0 < -\frac{1}{6}\lambda^2\rho_0, \qquad V_0 > 0. \tag{38}$$

If $V_0 > 0$, Eq (31) is negative and we have the period of deflation mentioned before. But $V_0 < -1/6\lambda^2\rho_0$ means that even in the region of vacuum domination ϵ can be positive and we return to the usual picture of inflation. However, having chosen a value of V_0 below this bound, Eq (26) becomes negative, so now we have a period of deflation translated to the epoch of field domination.

We can then choose values for V_0 below $-1/6\lambda^2\rho_0$ being in a period of deflation for the intermediate regime (Eq. (15) is negative) where Eqns. (19) and (18) still have real values because the approximation made in this section to find the upper limits of V_0, is a lower bound on Eqns (18) and (19). This is in fact redundant since we have also said that both terms in the potential (4) are of the same order, so the expansion is valid for the general case.

Once the choice of V_0 is done, we can solve numerically to find a value for ϕ_N in Eq (17), then it is introduced into Eq (24) and can be accepted or rejected depending on whether or not it fulfills the left-hand side. This value depends on ρ_0, the brane tension, and we present the results for different values of it considering that in order to have no incompatibilities with nucleosynthesis the brane tension must satisfy [42] $\rho_0 \geq 2\text{MeV}^4$, the authors take the number 1MeV^4, the difference arises due to the change of notation. We also check that the choice on ρ_0 satisfies the COBE constrain.

The results are shown in table 2.

4.5. Vacuum-Dominated Region Revisited

Following the argument in the preceeding paragraph, it would be possible to continue with the usual analysis to find the value of ϕ that finishes inflation, in the vacuum-dominated region provided V_0 is negative and calculate the number of e-foldings and then use Eq (24) in the region of vacuum domination to find a constrain on V_0. We find from equation (31) that

$$\sqrt{\kappa}\phi_e = \frac{1}{2\lambda}\ln\left(-\frac{4V_0}{\lambda^2\rho_0}\right), \tag{39}$$

and from Eq (14) that

$$\sqrt{\kappa}\phi_{60} \simeq -\frac{1}{\lambda}\ln\left[\left(-\frac{\lambda^2\rho_0}{4V_0}\right)^{1/2} - \frac{30\rho_0\lambda^2}{2V_0}\right]. \tag{40}$$

And finally from (24):

$$(-V_0)^{3/2} - 60\rho_0^{1/2}\lambda V_0 = \frac{\sqrt{75\pi}\delta_H\rho_0}{\kappa} \tag{41}$$

This equation can be solved numerically to give a value of V_0 that is in accordance with the COBE constrain for the perturbations. As we shall see later, this process will not be applied to the case of vacuum domination, since we find ϵ to be a decreasing function of time, therefore for this case inflation never ends. Consequently, it is not possible to find a value for the parameter V_0 satisfying this condition in the case of vacuum domination. So in fact, the value of ϵ corresponding to 1 indicates in this case the place where inflation starts to take place as from there onwards we will have that $\epsilon < 1$

4.6. Field Equations

In this subsection we solve the system of field equations. The numeric results are presented in the next section. Integrating Eqns.(11) for the potential (4) yields:

$$a(t) = a_0 \exp\left[-\frac{V_0}{\lambda^4 t\sqrt{3\kappa\rho_0^3}}(\lambda^4\kappa\rho_0 t^2 + 24e^{-2})\right] \tag{42}$$

and

$$\sqrt{\kappa}\phi(t) = -\frac{2}{\lambda} + \frac{1}{\lambda}\ln\left(\frac{48}{\lambda^4\kappa\rho_0 t^2}\right). \tag{43}$$

One can immediately see that the behavior of the field does not depend on the value of the parameter V_0, but only on λ whose value is given by Eq. (30). From this solution and its plot, one can check that indeed the field is a decreasing function of time, having the same bahaviour regardless of the regime it is in. The scale factor shows little dependence on the value of V_0 as shown in the plots. Following an increasing behavior for $V_0 > 0$. Figures 3, 4, 5 show that for $V_0 > 0$, ϵ is a growing positive function which corresponds to an inflationary stage as expected from the analysis of the previous sections. The acceleration factor is also shown, and one can observe that for the range used in the plots $\ddot{a}/a$ changes sign before ϵ reaches 1 in the same interval.

We have plotted the scale and acceleration factors as well as ϵ in Fig.6 for a negative value of V_0 The scale and acceleration factors change the sign of their slope at $\lambda^2(\kappa\rho_0)^{1/2}t = 1.8$ which is the same value as that from Eq. (33) indicating the start of the vacuum-dominated regime. So one ends with a stage of inflation after deflation in the other two regimes. We find that indeed ϵ is a positive decreasing function of time. This means that although there is a period of inflation, after deflation, it will never end and there is no meaning in calculating the value of the potential from Eq. (41) satisfying the COBE constrain since there is no value of the field corresponding to 60 e-folds before the end of inflation. The case of vacuum domination with a negative potential is not realistic for this model.

The bounds that the field needs to meet in order to have inflation for the intermediate regime are presented in table 3. The value of the parameter V_0 in the region of vacuum domination remains unconstrained, therefore it is not possible to give a bound on the value of the field for the onset of inflation.

4.7. Results

Before starting with the numeric results for the three regions just analyzed, we summarize what we have found so far. This is shown schematically in table 1.

Table 1. Results for the sign of the first slow-roll parameter ϵ according to the choice of V_0 for the three different regions of potential (4).

Region	$V_0 > 0$	dynamics	$V_0 < -\frac{1}{6}\lambda^2\rho_0$	dynamics
ϕ-dominated	$\epsilon > 0$, ϕ_e real	inflation	$\epsilon < 0$	deflation
Intermediate	$\epsilon > 0$, ϕ_e real	inflation	$\epsilon < 0$	deflation
Vacuum-dominated	$\epsilon < 0$	deflation	$\epsilon > 0$, ϕ_e real	inflation

Table 2. Results for V_0 in the intermediate regime, for three different values of ρ_0.

$\rho_0 \times 10^6$, (eV^4)	$\lambda \times 10^{14}$	$V_0 \times 10^{34}$, (eV^4)	$\phi_e \times 10^{13}$, (eV)	$\phi_{60} \times 10^{13}$, (eV)	δ_H, $\times 10^{-5}$
		1.4	1.5	2.7	1.9
2	8.2	1.1	1.6	2.8	1.9
		0.9	1.6	2.8	1.9
		2.2	1.7	3.1	1.9
4	7.4	1.8	1.8	3.1	1.9
		1.4	1.9	3.2	1.9
		2.8	1.9	3.3	1.9
6	6.9	2.4	1.9	3.3	1.9
		1.9	1.9	3.4	1.9

Despite the fact that $V_0 > 0$ gives a positive value for Eq. (26), it does not correspond to the region of field domination. So one has to employ the value of $V_0 > 0$ in Eqns. (15) and (18) not in (27). That is, in the intermediate regime. We have checked that $V_0 > 0$ does not meet the condition (21) even for very small values of V_0 compared to unity. Instead, the smaller this parameter is, the closer to 2 is Eq. (21). So we are left with only two regions where we have inflation. The intermediate regime for $V_0 > 0$, and the region of vacuum domination for $V_0 < -1/6\lambda^2\rho_0^2$.

We solved numerically the corresponding equations in the intermediate regime and found the values shown in table 2. For this, we have taken that $\kappa \simeq 25/m_{Pl}^2$ [42].

Table 2 shows 3 values of the potential that are in good agreement with the value of the density contrast δ_H. The numbers that appear in the third column correspond to $1/100\lambda^2\rho_0$, $1/120\lambda^2\rho_0$ and $1/150\lambda^2\rho_0$ respectively. Bigger values in the denominators seem to lead the decimals in the density contrast closer to 1.9×10^{-5}. We keep these numbers as a good approximation to the ideal value of the potential.

Table 3. Bounds for the scalar field ϕ multiplied by the Planck mass, for V_0 positive in eV units.

$\rho_0 \times 10^6$, (eV^4)	$\lambda \times 10^{14}$	V_0, (eV^4)	$\phi > \times 10^{-15} \times m_{Pl}$, (eV)
2	8.2	$\frac{1}{100}\rho_0\lambda^2$	1.3
		$\frac{1}{120}\rho_0\lambda^2$	1.3
		$\frac{1}{150}\rho_0\lambda^2$	1.4
4	7.4	$\frac{1}{100}\rho_0\lambda^2$	1.4
		$\frac{1}{120}\rho_0\lambda^2$	1.5
		$\frac{1}{150}\rho_0\lambda^2$	1.5
6	6.9	$\frac{1}{100}\rho_0\lambda^2$	1.5
		$\frac{1}{120}\rho_0\lambda^2$	1.6
		$\frac{1}{150}\rho_0\lambda^2$	1.7

5. The Two Fields Analysis

In this section we perform the analysis of the dynamics using only the $\cosh \sim \exp$ approximation. The complete field equations now read

$$H^2 = \frac{\kappa}{3}\left(\frac{1}{2}\dot{\phi}^2 + \frac{1}{2}\dot{\psi}^2 e^{\lambda\sqrt{\kappa}\phi} + V_\phi + e^{\lambda\sqrt{\kappa}\phi}V_\psi + \rho_\gamma e^{\alpha\sqrt{\kappa}\phi} + \rho_{V_0}\right), \tag{44a}$$

$$\ddot{\phi} + 3H\dot{\phi} + \frac{dV_\phi}{d\phi} = \lambda\sqrt{\kappa}\,e^{\lambda\sqrt{\kappa}\phi}\left(\frac{1}{2}\dot{\psi}^2 - V_\psi\right)\alpha\sqrt{\kappa}\,e^{\alpha\sqrt{\kappa}\phi}\rho_\gamma \tag{44b}$$

$$\ddot{\psi} + 3H\dot{\psi} + \frac{dV_\psi}{d\psi} = -\lambda\sqrt{\kappa}\,\dot{\phi}\,\dot{\psi}, \tag{44c}$$

$$\dot{\rho}_\gamma + 3H\rho_\gamma = 0, \tag{44d}$$

$$\dot{\rho}_{V_0} = 0, \tag{44e}$$

where we have set

$$V_\phi = \frac{1}{2}V_0 e^{\kappa\lambda\phi}$$
$$V_\psi = \frac{1}{2}V_0\kappa^2\psi_0^2\psi^2$$
$$\rho_{V_0} = V_0 \tag{45}$$

a dot stands for a derivative with respect to the cosmological time and H is the Hubble parameter $H = \dot{a}/a$. The slow-roll parameter $\epsilon = -\frac{\dot{H}}{H^2}$ can be obtained by differentiating

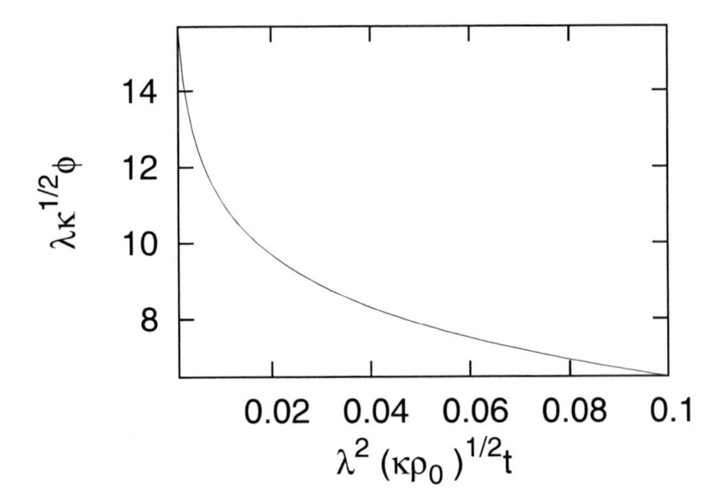

Figure 2. The behavior for the scalar field during inflation.

equation (44a) with respect to t. One arrives at

$$\dot{H} = -\frac{\kappa^2}{2}\left(\dot{\phi}^2 + \dot{\psi}^2 + \gamma\rho_\gamma\right) \tag{46}$$

We perform the following definitions for the dynamical variables

$$x = \frac{\kappa}{\sqrt{6}}\frac{\dot{\phi}}{H} \quad u = \frac{\kappa}{\sqrt{6}}\frac{\sqrt{V_\phi}}{H} \tag{47a}$$

$$A = \frac{\kappa}{\sqrt{6}}\frac{\dot{\psi}}{H}e^{\frac{1}{2}\lambda\kappa\phi} \quad w = \frac{\kappa}{\sqrt{6}}\frac{\sqrt{V_\psi}}{H}e^{\frac{1}{2}\lambda\kappa\phi} \tag{47b}$$

$$y = \frac{\kappa}{\sqrt{6}}\frac{\sqrt{\rho_\gamma}}{H}e^{\frac{1}{2}\alpha\kappa\phi} \quad l = \frac{\kappa}{\sqrt{6}}\frac{\sqrt{V_0}}{H} \tag{47c}$$

$$s = \kappa\psi_0\frac{\sqrt{V_0}}{H} \tag{47d}$$

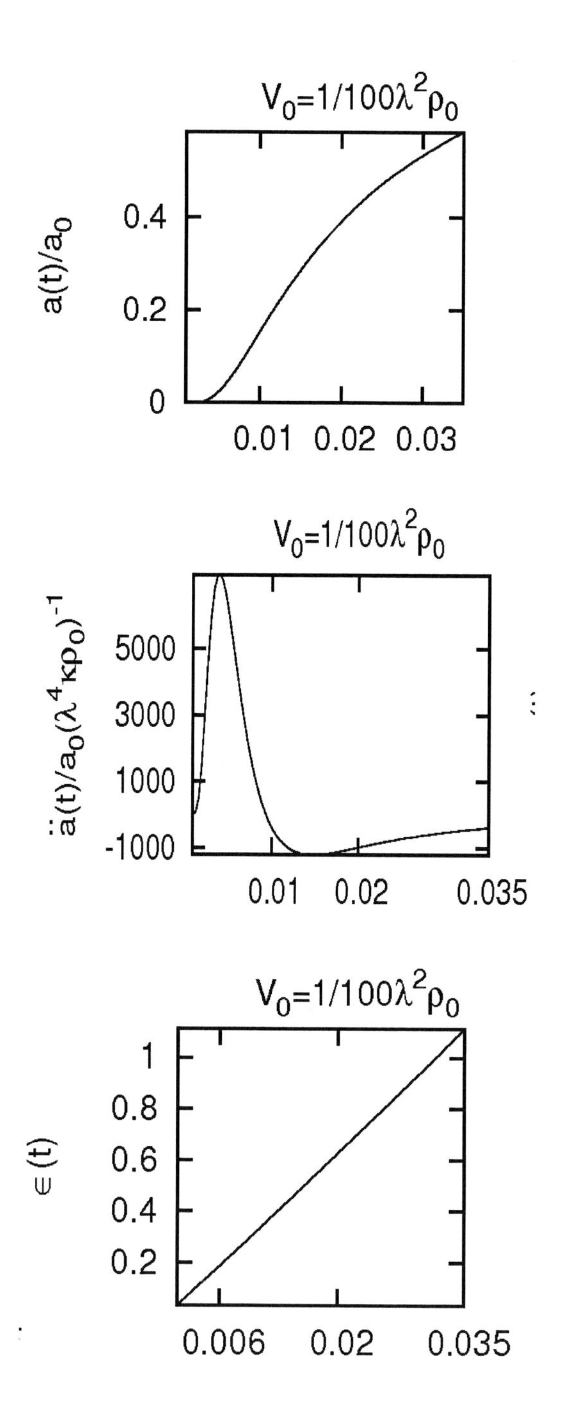

Figure 3. On the left hand side the scale factor shows an inflationary behavior but the acceleration factor, on the center, grows and decreases in the same interval. On the right hand side we plot the inflationary parameter ϵ

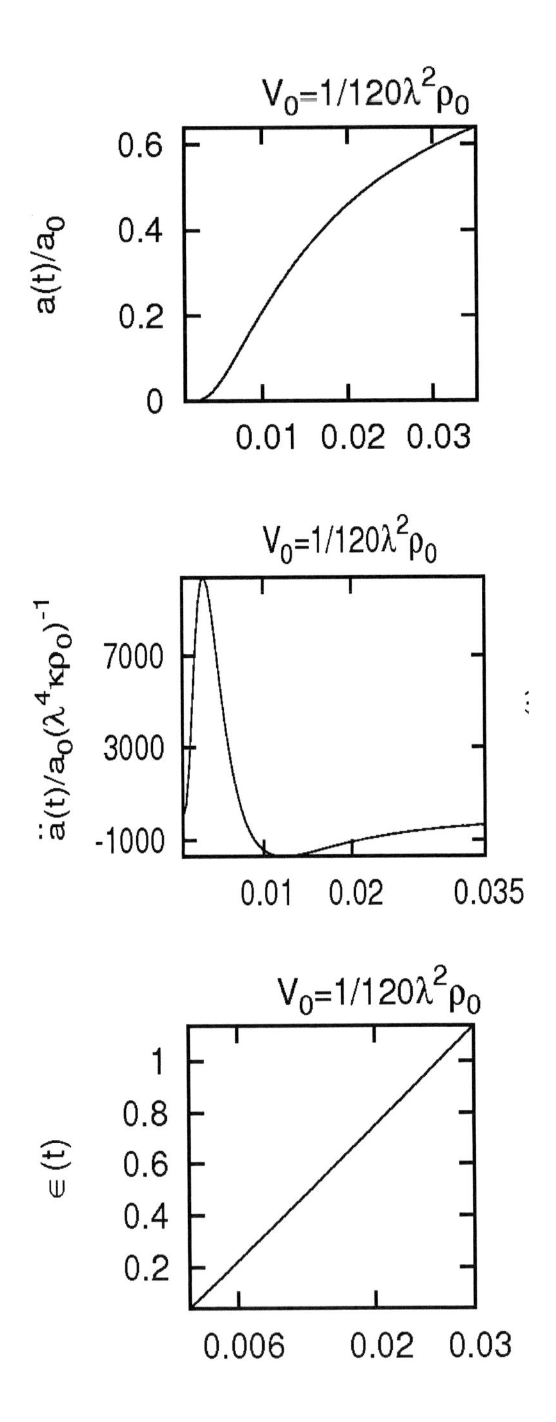

Figure 4. In this case the scale factor (lhs) increases more rapidly than the previous case and the acceleration factor (center) reaches higher numbers within the same interval. On the rhs, again we plot the ϵ parameter.

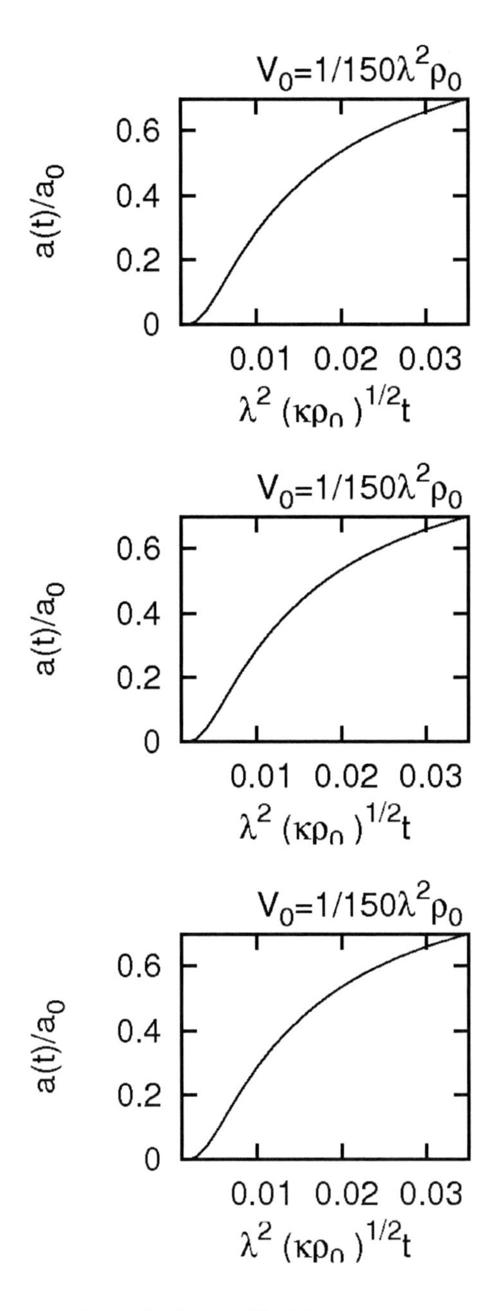

Figure 5. Again in this case the scale factor (lhs) increases more rapidly than in the first case and the acceleration factor (center) reaches much higher numbers within the same interval. On the rhs, again we plot the ϵ parameter.

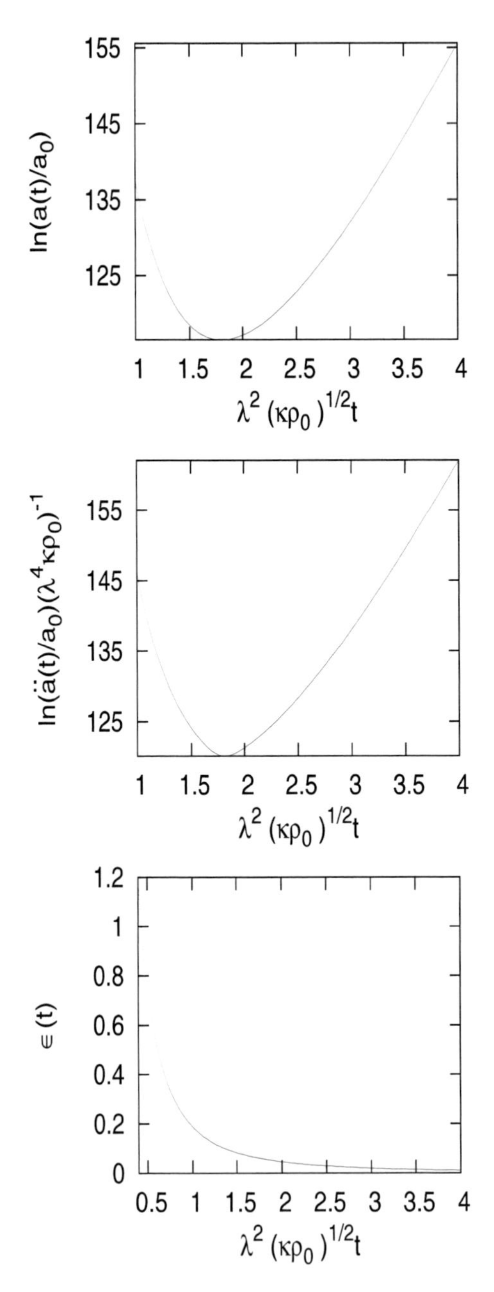

Figure 6. The plot of the scale factor, the acceleration factor and ϵ for V_0 negative, here we use $V_0 = -56\lambda^2\rho_0$

with this variables the field equations reduce to

$$x' = -3\,x + \frac{\lambda}{2}u^2 - \sqrt{\frac{3}{2}}\left(\lambda(w^2 - A^2) + \alpha y^2\right) + \frac{3}{2}\Pi\,x, \tag{48a}$$

$$A' = -3\,A - w\,s - \sqrt{\frac{3}{2}}\lambda\,A\,x + \frac{3}{2}\Pi\,A, \tag{48b}$$

$$u' = \frac{\lambda}{2}\,u\,x + \frac{3}{2}\Pi\,u, \tag{48c}$$

$$w' = A\,s + \lambda\sqrt{\frac{3}{2}}w\,x + \frac{3}{2}\,\Pi\,w \tag{48d}$$

$$y' = \frac{3}{2}\left(\Pi - \gamma + \alpha\sqrt{\frac{2}{3}}x\right)y, \tag{48e}$$

$$l' = \frac{3}{2}\Pi\,l, \tag{48f}$$

$$s' = \frac{3}{2}\Pi\,s, \tag{48g}$$

where now prime stands for the derivative with respect to the number of e-foldings $N = \ln(a)$. The quantity Π is related to the slow-roll parameter ϵ by $\Pi = \frac{2}{3}\epsilon$ and is defined as

$$\Pi = 2\,x^2 + 2\,A^2 + \gamma y^2 \tag{49}$$

The Friedman equation (44a) becomes a constriction for the variables such that

$$F = x^2 + A^2 + u^2 + w^2 + y^2 + l^2 = 1. \tag{50}$$

In order to see the dynamical evolution, we analyze the stability of system (48) using the theorem of Hartman-Grobman. We first perform a small perturbation $\delta\vec{x}$ of the system around of the minima $\vec{x} = \vec{x_c} + \delta\vec{x}$, in order to analyze the stability of these points. Doing so, the perturbation fulfills the relation

$$\delta\vec{x'} = \mathcal{M}\delta\vec{x}, \tag{51}$$

The face space we are interested in is defined by the vector $\vec{x} = (x, u, A, w, l, s)$. In this case the matrix $\mathcal{M}$ is given by

$$\begin{bmatrix} -3\,A_0{}^2+3-9\,x_0{}^2 & 0 & -\sqrt{6}\left(\sqrt{6}x_0+\lambda\right)A_0 & \sqrt{6}\lambda\,w_0 & 0 & 0 \\ -\frac{1}{2}\,u_0\,(\lambda+12\,x_0) & -3\,x_0{}^2-\frac{1}{2}\,\lambda\,x_0-3\,A_0{}^2 & -6\,A_0\,u_0 & 0 & 0 & 0 \\ \sqrt{\frac{3}{2}}\left(-2\,\sqrt{6}x_0+\lambda\right)A_0 & 0 & \sqrt{\frac{3}{2}}\lambda\,x_0+3-9\,A_0{}^2-3\,x_0{}^2 & 0 & 0 & 0 \\ -\sqrt{\frac{3}{2}}\left(2\,\sqrt{6}x_0+\lambda\right)w_0 & 0 & -s_0-6\,A_0\,w_0 & -\sqrt{\frac{3}{2}}\lambda\,x_0-3\,x_0{}^2-3\,A_0{}^2 & 0 & -A_0 \\ -6\,x_0\,l0 & 0 & -6\,A_0\,l_0 & 0 & -3\,x_0{}^2-3\,A_0{}^2 & 0 \\ -6\,x_0\,s_0 & 0 & -6\,A_0\,s0 & 0 & 0 & -3\,x_0{}^2-3\,A_0{}^2 \end{bmatrix}$$

The critical points $\vec{x_c} = (x_0, u_0, A_0, w_0, l_0, s_0)$ which fulfill the Friedman constriction (50) are then

- $(0, 0, 0, 0, l, s) \rightarrow V_0$ dominance

- $(\pm 1, 0, 0, 0, 0, 0) \rightarrow$ dilaton's kinetic part dominance.
- $\frac{1}{\sqrt{6}}(-\lambda, 0, 0, \sqrt{6-\lambda^2}, 0, 0) \rightarrow$ dilaton's kinetic part plus axion's potential dominance.
- $\frac{1}{2}(-\frac{\sqrt{6}}{\lambda}, 0, \frac{\sqrt{2\lambda^2-6}}{\lambda}, \sqrt{2}, 0, 0) \rightarrow$ dilaton's kinetic part plus axion dominance.
- $\frac{1}{6}(-\lambda, \frac{1}{\sqrt{2}}, 0, 0, 0, 0) \rightarrow$ dilaton dominance, with $\lambda = \pm\sqrt{\frac{71}{2}}$.
- $\frac{1}{5}(\frac{6(1-\sqrt{6})}{\lambda}, \sqrt{5(6-\sqrt{6})}, \frac{\sqrt{36(2\sqrt{6}-7)-5\lambda^2(1-\sqrt{6})}}{\lambda}, 0, 0, 0) \rightarrow$ dilaton plus axion's kinetic part dominance.

In order to obtain the stability of this points, we substitute them into the matrix (51) and obtain their eigenvalues, we arrive at

- V_0 dominance, the eigenvalue of $\mathcal{M}$ is

$$(-3, -3, 0, 0, 0, 0)$$

  This is an attracting wall.

- Dilaton's kinetic part dominance, the eigenvalue of $\mathcal{M}$ is

$$(3 + \sqrt{\frac{3}{2}}\,\lambda, -\sqrt{\frac{3}{2}}\,\lambda, 3 + \frac{\lambda}{2}, 3, 3, 6)$$

  This point is a repelling focus for $-3\sqrt{\frac{3}{2}} < \lambda < 0$, but it is a saddle point otherwise.

- Dilaton's kinetic part plus axion's potential dominance, the eigenvalue of $\mathcal{M}$ is

$$(\frac{1}{2}\,\lambda^2 - 3, \lambda^2, \frac{1}{2}\,\lambda^2, \frac{1}{2}\,\lambda^2, \lambda^2 - 3, \frac{1}{2}\lambda^2\left(1 - \frac{\sqrt{6}}{12}\right))$$

  This is a repelling focus for $\lambda > \pm\sqrt{6}$, but the critical point becomes then imaginary. If $\lambda \neq 0$ but $-\sqrt{6} < \lambda < \sqrt{6}$ this is a saddle point and it is an attractor wall if $\lambda = 0$, then the point becomes an axion's potential dominance.

- Dilaton's kinetic part plus axion dominance, the eigenvalue of $\mathcal{M}$ is

$$(-\frac{3}{4} - \frac{1}{4}\,\sqrt{153 - 48\,\lambda^2}, -\frac{3}{4} + \frac{1}{4}\,\sqrt{153 - 48\,\lambda^2}, -\frac{\sqrt{6}}{4} + \frac{3}{2}, \frac{3}{2}, \frac{3}{2}, 3)$$

  This is a saddle point.

- Dilaton dominance, the eigenvalue of $\mathcal{M}$ is

$$(\frac{1}{24}\,(1 - 71\sqrt{6}), \frac{71}{24}(-1 + \sqrt{6}), \frac{71}{24}, \frac{71}{24}, \frac{47}{8}, 0)$$

  This is a saddle point.

- Dilaton plus axion's kinetic part dominance, is a saddle point.

From these results we can see that only the constant part of the dilaton V_0 can dominate the evolution of the system. In all the other cases the system contains only saddle or source points. In Fig 7 we see that the slow-roll parameter ϵ remains small during the evolution, we infer that during this time the system undergoes an inflationary period of evolution. In this figure the system starts with initial conditions where the constant V_0 dominates the evolution, but this point on the phase space is stable, thus the system evolves into an eternal period of inflation. We plot the evolution of each field and the phase spaces (A, w, x) and (x, l) to see the evolution of the fields on this period. In Fig 8 which represents the cases where the axion's kinetic part and the dilaton's potential part dominate the evolution, we see that the slow-roll parameter ϵ is always small, we could expect an inflationary period of evolution from this initial condition. Fig 9 corresponds to the dilaton's kinetic part plus axion's potential dominance with $\lambda = 1$. Here the slow roll parameter ϵ oscillates very hard, we cannot expect an inflationary period with this initial conditions.

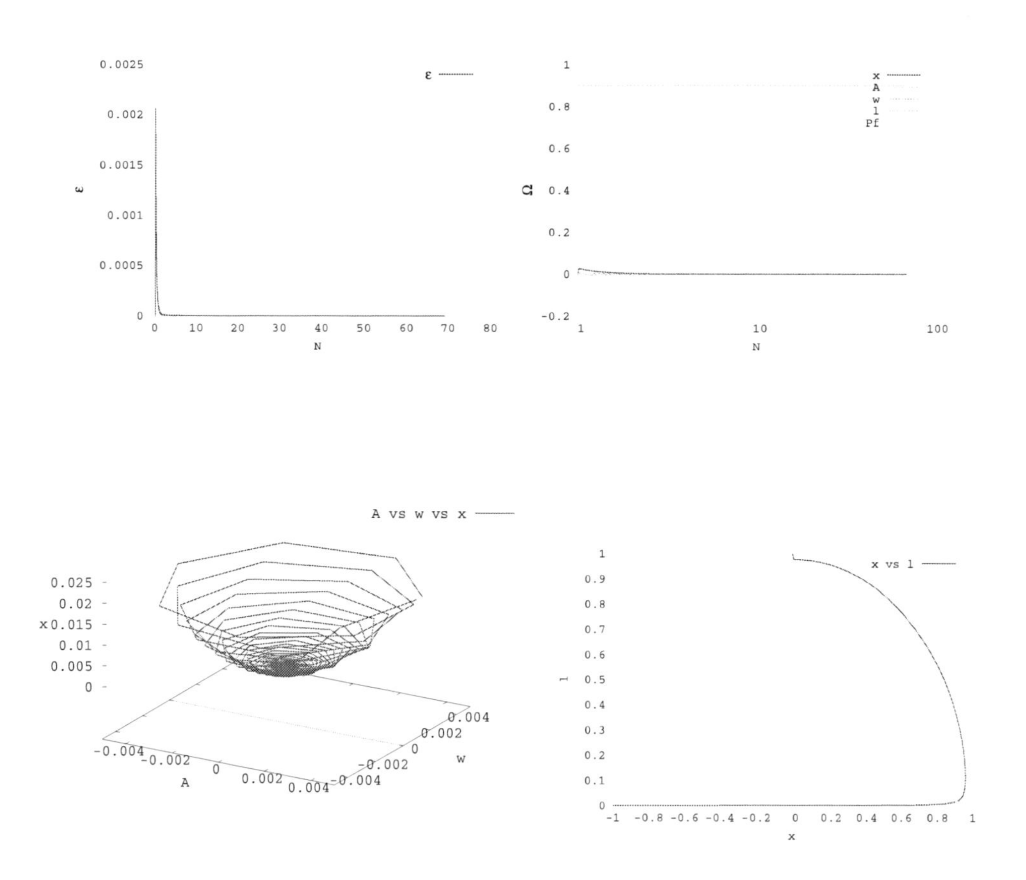

Figure 7. We plot ϵ (left upper panel) the evolution of the fields (right upper panel), the phase spaces (A, w, x) (left lower panel) and (x, l) (right lower panel) for the initial conditions $(0, 0, 0, 0, 1, 100)$. This initial conditions correspond to the V_0 dominance, observe how the (A, w, x) fields oscillate arround the $(0, 0, 0)$ and the $l = 1$ dominates the evolution. The slow roll parameter ϵ is allway very small.

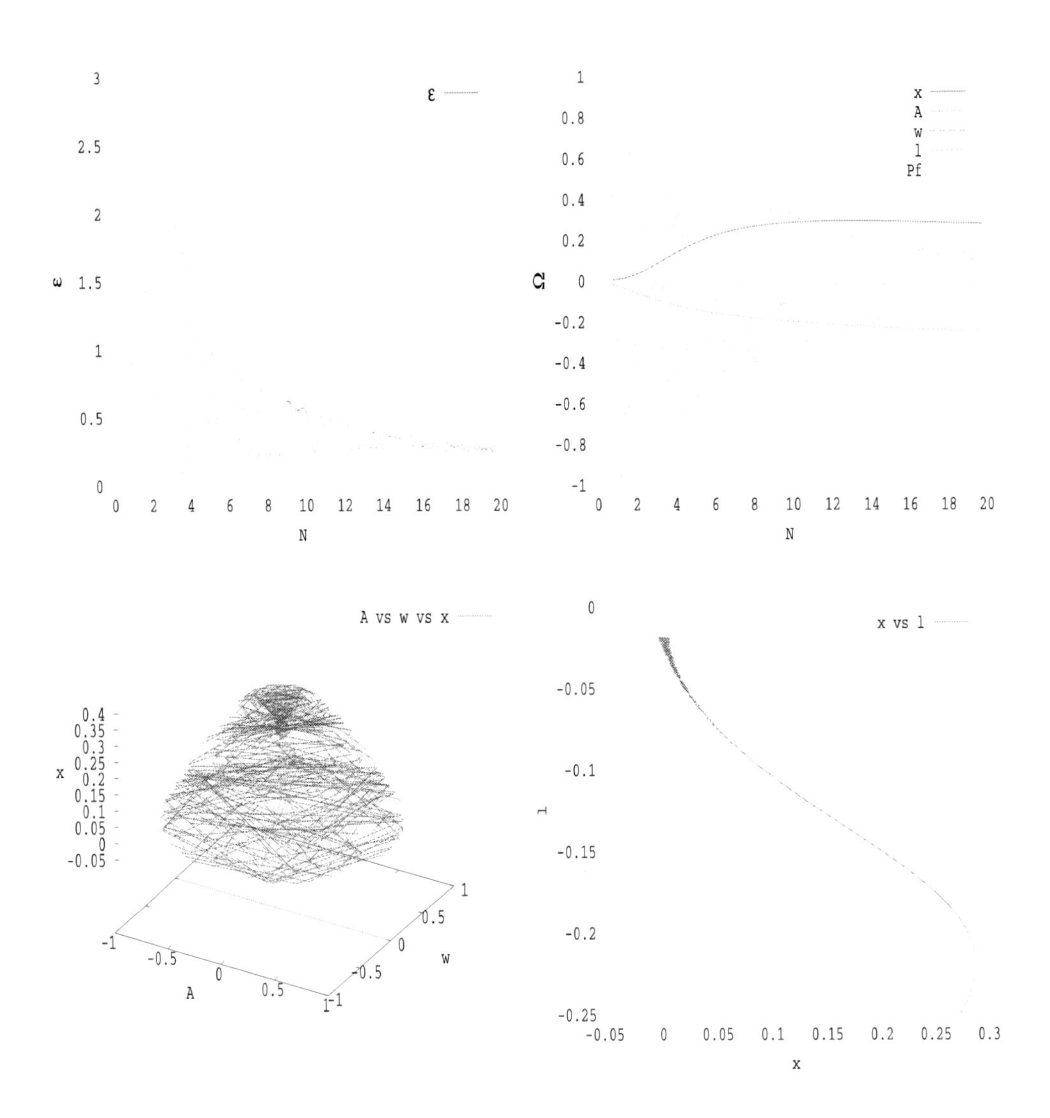

Figure 8. We plot ϵ (left upper panel) the evolution of the fields (right upper panel), the phase space (A, w, x) (left lower panel) and the phase space (x, l) (right lower panel) for the initial conditions $(0, 0.01, 0, 0.99, 0, 100)$. This initial condition corresponds to a saddle point with $\lambda = 1$. The fields oscillate, but ϵ goes oscillating to a fix point less than $1/2$. Here non of the fields dominate the evolution, there is a dynamical equilibrion of all of them.

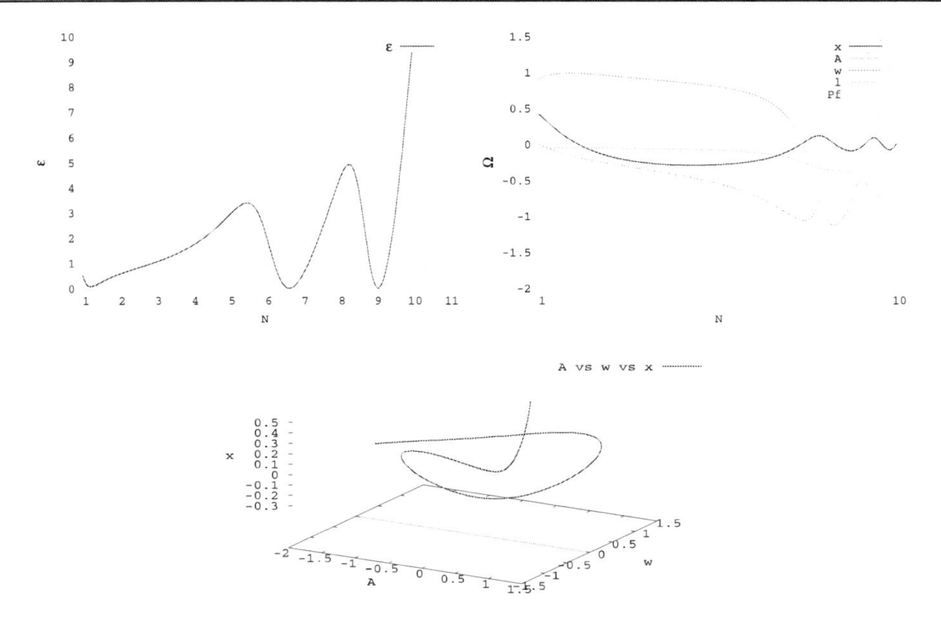

Figure 9. We plot ϵ (left upper panel) the evolution of the fields (right upper panel), the phase space (A, w, x) (left lower panel) and the phase space (x, l) (right lower panel) for the initial conditions given by the dilaton's kinetic part plus axion's potential dominance with $\lambda = 1$

6. The Cosmology

Now, we proceed to describe the different epochs of the universe using this new interpretation. We can easily distinguish two behaviors of the scalar field potential: the exponential and the power laws. In the early universe the exponential behavior dominates the scalar fields potential. In this case we have the following analysis.
Inflation.- In this epoch, the scalar field potential can be written as

$$V = V_0 \exp\left(\lambda\sqrt{\kappa}\phi\right)\left(1 + \frac{1}{2}\,\kappa\psi_0{}^2\psi^2\right), \tag{52}$$

because the exponential dominates completely the scenario of the evolution of the dilaton potential. The distinctive feature during this period is that the presence of the fluxes generate a quadratic term in the Friedman equation. The scalar field density $\rho = 1/2\,\dot{\phi}^2 + 1/2\,\dot{\psi}^2\,e^{\tilde{\lambda}\sqrt{\kappa}\phi} + V$ appears quadratic in the field equations,

$$H^2 = \frac{\kappa}{3}\rho\left(1 + \frac{\rho}{\rho_0}\right). \tag{53}$$

Under this conditions it is known that these potentials are always inflationary in the presence of these fluxes [17]. Nevertheless, exponential potentials are inflationary without branes, in the traditional Friedman cosmology, only if $\lambda^2 < 2$ (see for example [18]). Therefore, if we suppose that $\lambda^2 > 2$, the dilaton potential (2) is not inflationary without the quadratic density term. Thus, as the universe inflates, the quadratic term becomes much more smaller than the linear term and we recover the Friedman equation $H^2 = \frac{\kappa}{3}\rho$, where the exponential potential is not inflationary anymore. For these values of λ this gives a natural graceful-exit to this scalar field potential [19]. It remains to study which is the influence of the axion potential to this epoch [20].

Densities Evolution.- The evolution of the densities is quite sensible to the initial conditions. Let us study the example of evolution shown in Fig.10. As in the ΛCDM model, here also the recombination period starts around the redshift 10^3. The first difference we find between ΛCDM model and the IIB superstrings theory is just between the redshifts 10^3 and 10^2, where the interaction between the dilaton and matter gives rise to oscillations of the densities. It is just in this epoch where we have to look for observations that can distinguish between these two models. In this epoch the scalar field is already small $\lambda\sqrt{\kappa}|\phi| < 4$ and approaches the minimum of the potential in $\phi = 0$. Thus, potential (2) starts to behave as a power low potential, simulating a type ϕ^2 field. Therefore it is not surprising that this potential mimics very well the dark matter behavior. In the ΛCDM model, dark matter is modeled as dust and it is well known that power low potentials mimics dust fluids as they oscillate around the minimum of the potential [21]. In figures Fig.10 and Fig.11 this behavior is confirmed.

Nevertheless, for redshifts bigger than $1/a - 1 = z \sim 10^3$, there are remarkable differences between the superstring model and the CDM one. The interaction of the dilaton field with matter provokes to be very difficult that radiation dominates the universe, thus big bang nucleosynthesis never takes place, at lest in a similar way as in the CDM paradigm. Let us explain this point. The dilaton field interacts with matter through the factor $e^{\tilde{\alpha}(\Phi-\Phi^{(0)})}\, F^2 = e^{\alpha\sqrt{\kappa}\phi}\, F^2$, being F the field strength of the matter contents. Thus, Lagrangian for the superstrings system is

$$\mathcal{L} = \sqrt{-g}\left(R - \mathcal{L}_\phi - e^{\tilde{\lambda}\sqrt{\kappa}\phi}\mathcal{L}_\psi - e^{\alpha\sqrt{\kappa}\phi}\mathcal{L}_{matter}\right), \tag{54}$$

where we have differentiated the scalar field potential coupling constant λ from the axion-dilaton coupling constant $\tilde{\lambda}$ in order to generalized and clarify the cosmology of the system. In (1) both are the same $\lambda = \tilde{\lambda}$. The individual Lagrangians for the dilaton and axion fields respectively are,

$$\mathcal{L}_\phi = \frac{1}{2}\partial^\sigma\phi\partial_\sigma\phi + V_\phi, \quad \mathcal{L}_\psi = \frac{1}{2}\partial^\sigma\psi\partial_\sigma\psi + V_\psi. \tag{55}$$

Thus, in a flat Friedman-Robertson-Walker space-time the cosmological field equations are given by

$$\begin{aligned} H^2 &= \frac{\kappa}{3}\left(\frac{1}{2}\dot{\phi}^2 + \frac{1}{2}\dot{\psi}^2 e^{\tilde{\lambda}\sqrt{\kappa}\phi} + V_\phi + e^{\tilde{\lambda}\sqrt{\kappa}\phi}V_\psi\right. \\ &+ \left.(\rho_b + \rho_{rad})\, e^{\alpha\sqrt{\kappa}\phi} + \rho_L\right), \end{aligned} \tag{56a}$$

$$\begin{aligned}
\ddot{\phi} &+ 3\,H\dot{\phi} + \frac{d\,V_\phi}{d\phi} = \tilde{\lambda}\sqrt{\kappa}\,e^{\tilde{\lambda}\sqrt{\kappa}\phi}\left(\frac{1}{2}\dot{\psi}^2 - V_\psi\right) \\
&- \alpha\sqrt{\kappa}\,e^{\alpha\sqrt{\kappa}\phi}(\rho_b + \rho_r) &(56b)\\
\ddot{\psi} &+ 3\,H\dot{\psi} + \frac{d\,V_\psi}{d\psi} = -\tilde{\lambda}\sqrt{\kappa}\,\dot{\phi}\,\dot{\psi}, &(56c)\\
\dot{\rho}_b &+ 3\,H\,\rho_b = 0, &(56d)\\
\dot{\rho}_{rad} &+ 4\,H\,\rho_{rad} = 0, &(56e)\\
\dot{\rho}_L &+ 3\,\gamma_{DE}\,H\,\rho_L = 0, &(56f)
\end{aligned}$$

where the dot stands for the derivative with respect to the cosmological time and H is the Hubble parameter $H = \dot{a}/a$. In order to analyze the behavior of this cosmology, we transform equations (56a)-(56f) using new variables defined by

$$\begin{aligned}
x &= \frac{\sqrt{\kappa}}{\sqrt{6}}\frac{\dot{\phi}}{H},\quad A = \frac{\sqrt{\kappa}}{\sqrt{6}}\frac{\dot{\psi}}{H}e^{\frac{1}{2}\tilde{\lambda}\sqrt{\kappa}\phi}, &(57a)\\
y &= \frac{\sqrt{\kappa}}{\sqrt{3}}\frac{\sqrt{\rho_b}}{H}e^{\frac{1}{2}\alpha\sqrt{\kappa}\phi},\quad z = \frac{\sqrt{\kappa}}{\sqrt{3}}\frac{\sqrt{\rho_{rad}}}{H}e^{\frac{1}{2}\alpha\sqrt{\kappa}\phi}, &(57b)\\
u &= \frac{\sqrt{\kappa}}{\sqrt{3}}\frac{\sqrt{V_\phi}}{H},\quad v = \frac{\sqrt{\kappa}}{\sqrt{3}}\frac{\sqrt{V_2}}{H}, &(57c)\\
l &= \frac{\sqrt{\kappa}}{\sqrt{3}}\frac{\sqrt{\rho_L}}{H},\quad w = \frac{\sqrt{\kappa}}{\sqrt{3}}\frac{\sqrt{V_\psi}}{H}e^{\frac{1}{2}\tilde{\lambda}\sqrt{\kappa}\phi}, &(57d)
\end{aligned}$$

where we have used the definition of the potentials $V_\phi = 2\,V_0\sinh(1/2\,\sqrt{\kappa}\lambda\phi)^2$, $V_2 = 2\,V_0\cosh(1/2\,\sqrt{\kappa}\lambda\phi)^2$ and $V_\psi = \frac{1}{2}V_0\kappa{\psi_0}^2\psi^2$ such that $V = V_\phi + V_\psi e^{\tilde{\lambda}\sqrt{\kappa}\phi}$ is the total scalar field potential. With these definitions equations (56a)-(56f) transform into

$$\begin{aligned}
x' &= -3\,x - \sqrt{\frac{3}{2}}\left(\lambda uv + \alpha(y^2+z^2) + \tilde{\lambda}(w^2 - A^2)\right) \\
&+ \frac{3}{2}\Pi\,x, &(58a)\\
A' &= -3\,A - \sqrt{3}\,\frac{\psi_0\sqrt{V_0}}{\sqrt{\rho_L}}w\,l - \sqrt{\frac{3}{2}}\tilde{\lambda}\,A\,x + \frac{3}{2}\Pi\,A, &(58b)
\end{aligned}$$

$$
\begin{aligned}
y' &= \frac{3}{2}\left(\Pi - 1 + \alpha\sqrt{\frac{2}{3}}x\right) y, &(58c)\\
z' &= \frac{3}{2}\left(\Pi - \frac{4}{3} + \alpha\sqrt{\frac{2}{3}}x\right) z, &(58d)\\
u' &= \sqrt{\frac{3}{2}}\lambda\, v\, x + \frac{3}{2}\Pi\, u, &(58e)\\
v' &= \sqrt{\frac{3}{2}}\lambda\, u\, x + \frac{3}{2}\Pi\, v, &(58f)\\
l' &= \frac{3}{2}\left(\Pi - \gamma_{DE}\right) l, &(58g)\\
w' &= \sqrt{3}\frac{\psi_0\sqrt{V_0}}{\sqrt{\rho_L}} A\, l + \tilde{\lambda}\sqrt{\frac{3}{2}}w\, x + \frac{3}{2}\Pi\, w &(58h)
\end{aligned}
$$

where now prime stands for the derivative with respect to the N-foldings parameter $N = \ln(a)$. The quantity Π is defined as

$$
\Pi = 2\,x^2 + 2\,A^2 + y^2 + \frac{4}{3}\,z^2 \tag{59}
$$

The Friedman equation (56a) becomes a constriction of the variables such that

$$
x^2 + A^2 + y^2 + z^2 + u^2 + l^2 + w^2 = 1. \tag{60}
$$

The density rate quantities $\Omega_x = \rho_x/\rho_{critic}$ can be obtained using the variables (57a) - (57d), one arrives at

$$
\begin{aligned}
\Omega_{DM} &= x^2 + u^2,\\
\Omega_{DE} &= l^2,\\
\Omega_b &= y^2,\\
\Omega_{rad} &= z^2,\\
\Omega_A &= 1 - x^2 - u^2 - y^2 - z^2 - l^2, &(61)
\end{aligned}
$$

where Ω_{DM}, Ω_{DE}, Ω_b, Ω_{rad} and Ω_A respectively are the density rates for the dark matter (dilaton field), dark energy (cosmological constant), baryons, radiation and axion field. For the definition of this last one we have used the constriction (60). Equations (58a)-(58h) are now a dynamical system. The complete analysis of this system will be given elsewhere [22], but the main results are the following. 1.– The system contains many critical points, some of them are atractors with dark matter dominance, other with dark energy dominance. 2.– The system depends strongly on the initial conditions. One example of the evolution of the densities is plotted in Fig.10 and Fig.11, where we show that the densities behave in a very similar way as the corresponding ones of the ΛCDM model before redshifts 10^2, which seem to be a generic behavior. The free constants λ, α and $\tilde{\lambda}$ are given in each figure. On the other side, we can see that after redshifts $z \sim 10^3$ one finds that $|\phi| < 0.04\, m_{Planck}$

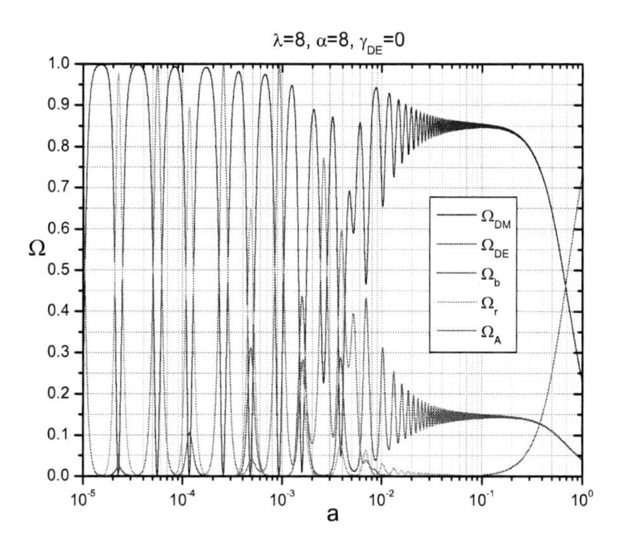

Figure 10. Plot of the dynamics of the Ω's in the type IIB superstring theory with fluxes. Observe how this theory predicts a similar behavior of the matter content of the universe as the ΛCDM model. Here, the initial values of the dynamical variables at redshift $a = 1$ are: $x = 0$, $A = 0$, $u = \sqrt{0.23}$, $v = 1000$, $\Omega_{DE} = 0.7299$, $\Omega_b = 0.04$, $\Omega_{rad} = 4 \times 10^{-5}$, and w is determined by the Friedman restriction. The values for the constants are $\alpha = \lambda = 8$, $\Psi_0\sqrt{V_0/\rho_L} = 40$, $\tilde{\lambda} = 0.1$. In all figures, the integration was made using the Adams-Badsforth-Moulton algorithm (variable step size). Each curve contains over 1×10^6 points.

and oscillating goes to zero, such that its exponential is bounded $0.01 < e^{-\lambda\sqrt{\kappa}\phi} < 1$ see Fig. 12. In other words, it takes the exponential more than 13 Giga years to change from 0.01 to 1.

However, there is one fact that takes our attention in Fig10. We see from the behavior of the densities on the early universe after redshifts $\sim 10^3$, that radiation does not dominates the rest of the densities as it is required for big bang nucleosynthesis. This fact can also be seen as follows. In a radiation dominated universe we might set $l = y = u = v = w = A = 0$, in that case we can see by inspection of (58a)-(58h) that there is no way that radiation remains as a dominant component of the system. The situations radically change if we put $\tilde{\lambda} = 0$ in system (58a)-(58h), in this case radiation has no problems to be dominant somewhere. In order to show how the dilaton and axion interaction with matter work, we study the particular case $\tilde{\lambda} = 0$ and let us artificially drop out the matter interaction from the dilaton equation (56b). In what follows we study this toy model. For this one it is convenient to change the variable w for $w = V_3$, with the definition of the potentials $V_3 = \sqrt{\kappa}\psi_0\psi$ such that $V = V_\phi^2 + 1/4(V_1 + V_2)^2 V_3^2$. Thus, equations (58a)-(58h) transform into the new

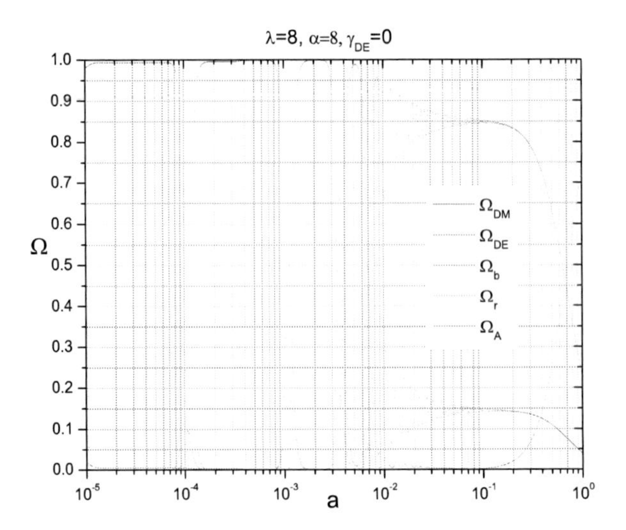

Figure 11. Plot of the dynamics of the Ω's in the type IIB superstring theory with fluxes. Initial values of the dynamical variables at redshift $a = 1$ are the same as in Fig 10. The values for the constants are $\alpha = 2 \times 10^{-5}$, $\lambda = 2.0$, $\Psi_0\sqrt{V_0/\rho_L} = 40$, $\tilde{\lambda} = 0$. Each curve contains over 5×10^5 points.

system

$$
\begin{aligned}
x' &= -3\,x - \sqrt{\frac{3}{2}}\left(\lambda\,uv + \frac{\alpha}{4}(v+u)^2w^2\right) + \frac{3}{2}\Pi\,x,\\
A' &= -3\,A - \sqrt{\frac{3}{2}}\psi_0\,\frac{1}{2}(v+u)^2w + \frac{3}{2}\Pi\,A,\\
y' &= \frac{3}{2}\left(\Pi - 1 + \alpha\sqrt{\frac{2}{3}}x\right)y,
\end{aligned}
$$

$$
\begin{aligned}
z' &= \frac{3}{2}\left(\Pi - \frac{4}{3} + \alpha\sqrt{\frac{2}{3}}x\right)z,\\
u' &= \sqrt{\frac{3}{2}}\lambda\,v\,x + \frac{3}{2}\Pi\,u,\\
v' &= \sqrt{\frac{3}{2}}\lambda\,u\,x + \frac{3}{2}\Pi\,v,\\
l' &= \frac{3}{2}\left(\Pi - \gamma_{DE}\right)l,\\
w' &= \sqrt{6}\psi_0\,A,
\end{aligned}
\qquad (62)
$$

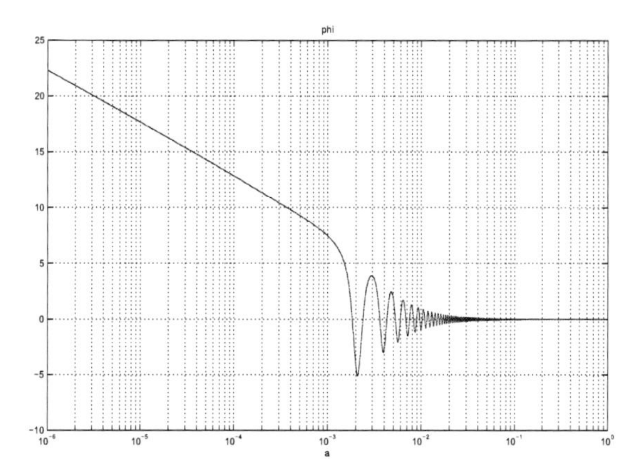

Figure 12. Plot of the behavior of the scalar field ϕ. The scalar field starts from big values and riches very fast its minimum where it starts to oscillate. We plot $\lambda\sqrt{\kappa}\phi$, for $\lambda = 20$.

The quantity Π is now defined as

$$\Pi = 2\,x^2 + 2\,A^2 + y^2 + \frac{4}{3}\,z^2 + \gamma_{DE}\,l^2 - \lambda\sqrt{\frac{2}{3}}\left(y^2 + z^2\right)x \tag{63}$$

and the new Friedman constriction (56a) reads

$$x^2 + A^2 + y^2 + z^2 + u^2 + l^2 + \frac{1}{4}\,(u + v)^2\,w^2 = 1. \tag{64}$$

The density quantities Ω_x now are

$$\begin{aligned} \Omega_{DM} &= x^2 + u^2, \\ \Omega_{DE} &= l^2, \\ \Omega_b &= y^2, \\ \Omega_{rad} &= z^2, \\ \Omega_A &= 1 - x^2 - u^2 - y^2 - z^2 - l^2, \end{aligned} \tag{65}$$

where we have used the constriction (64). Equations (62) are now a new dynamical system. The evolution of this one is shown in Fig13. From here we can see that now radiation dominates the early universe without problems and that the behavior of the densities is again very similar to the ΛCDM model but now for all redshifts. The only difference is at redshifts $10^2 < z < 10^3$, where the densities oscillate very hard. Unfortunately this time corresponds to the dark age, when the universe has no stars and there is nothing to observe which could give us some observational clue for this behavior.

Finally, if the set both coupling constant $\alpha = \tilde{\lambda} = 0$, we recover a very similar behavior of the densities to the ΛCDM model, this behavior es shown in Fig.14. Observe here that the densities have not oscillations any more, as in the ΛCDM model, supporting the idea that it is just the coupling between dilaton, axion and matter which makes difficult that the string theory reproduces the observed universe.

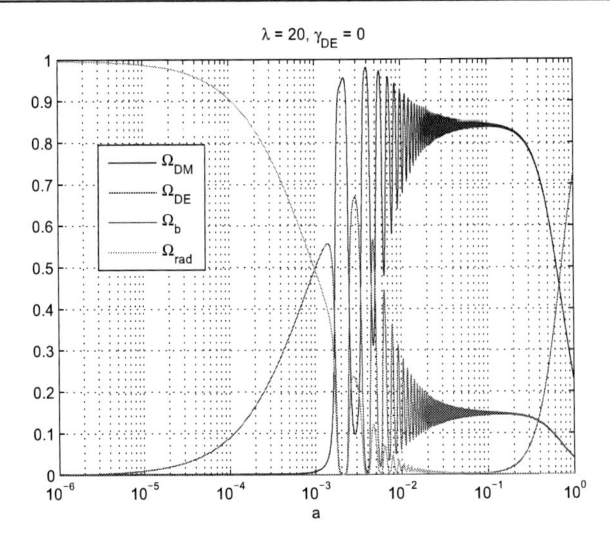

Figure 13. Plot of the dynamics of the Ω's in the type IIB superstring theory with fluxes. Observe how this theory predicts a similar behavior of the matter content of the universe as the ΛCDM model, even for redshifts beyond 10^3. Here radiation dominates the universe for values less than $a \sim 10^{-3}$ and big bang nucleosynthesis takes place as in the CDM model. Here $\lambda = \alpha = 20$, the initial values of the dynamical variables at redshift $a = 1$ are: $x = 0$, $A = 0$, $u = \sqrt{0.23}$, $v = 1000$, $\Omega_{DE} = 0.7299$, $\Omega_b = 0.04$, $\Omega_{rad} = 4 \times 10^{-5}$, and w is determined by the Friedman restriction. Each curve contains over 3×10^5 points.

Structure Formation.- As shown in figures Fig.10 and Fig.13 the axion field can be completely subdominant, but it can dominates the universe at early times as in Fig.11. At late times, $10^{-2} < a < 1$, the structure formation is determined by the dilaton field ϕ and its effective potential (2). In [23] it was shown that the scalar field fluctuations with a cosh potential follow the corresponding ones of the cold dark matter (CDM) model for the linear regime. There, it is shown that the field equations of the scalar field fluctuations can be written in terms of the ones of the ΛCDM model, in such a way that both models predict the same spectrum in the linear regime of fluctuations.

Galaxies Formation.- Other main difference between both models, the CDM and type IIB superstrings is just in the non-linear regime of fluctuations. Here numerical simulations show that the scalar field virialize very early [24], causing that in the superstring model galaxies form earlier than in the CDM paradigm. Furthermore, it has been shown that the scalar field does not have a cuspy central density profile [25]. Numerical and semi-analytic simulations have shown that the density profiles of oscillations (collapsed scalar fields) are almost flat in the center [26],[27],[24]. It has been also possible to compare high and low surface brightness galaxies with the scalar field model and the comparison shows that there is a concordance between the model and the observations, provided that the values of the parameters are just $V_0 \sim (3 \times 10^{-27} m_{Planck})^4$, $\lambda \sim 20$ [26]. With this values of V_0 and λ, the critical mass for collapse of the scalar field is just $10^{12} M_\odot$ [27], as it is expected for the halos of galaxies. These two features of the scalar field collapse might give

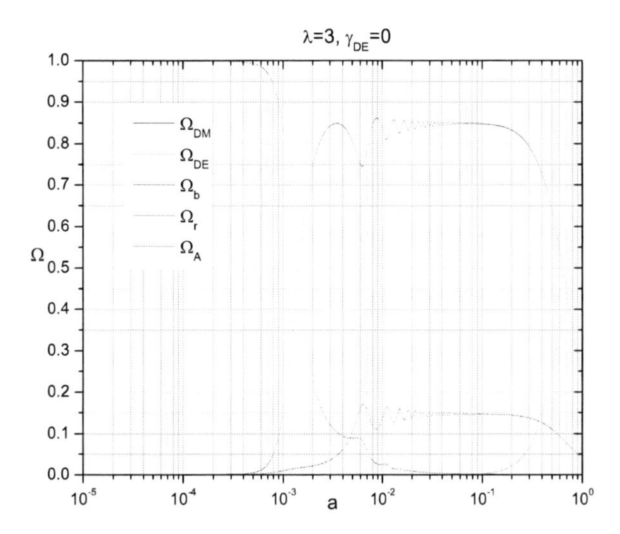

Figure 14. Plot of the dynamics of the Ω's in the cosh model. Here all the coupling constants of the superstrings model $\alpha = \tilde{\lambda} = 0$, $\lambda = 0.001$, the initial values of the dynamical variables at redshift $a = 1$ are the same as in Fig10. Observe how this theory predicts an extremely similar behavior of the matter content of the universe as the ΛCDM model, for all redshifts. Each curve contains over 3×10^5 points.

distinctive features to superstring theory. At the present time there is a controversy about the density profiles of the dark matter in the centers of the galaxies [28]. This model of the superstring theory predicts that the center of the galaxies contains an almost flat central density profile. We are aware that this result corresponds to the particular compactification $\mathbf{T}^6/\mathbb{Z}_2$, but it could be a general signature of string theory, in the sense that it could survive in a more realistic compactification (including branes and fluxes), that give rise to models that resemble the Standard Model. In this case, if the cuspy dark matter density profiles are observed or explained in some way, this model would be ruled out. But if these profiles are not observed, it would be an important astrophysical signature of string theory.

7. Conclusions

In the first part of this work we have seen the conditions that the parameters of the potential (4) have to fulfill in order to have early universe inflation. Addressing the question of whether the potential as quoted in [39] can give a unified picture between inflation and structure formation. We have done the first part of the analysis approximating the potential to be only a function of the dilaton field assuming the axion at the minimum and using the slow-roll approximtion. We find that the value of the parameter λ can be fixed in the region of field domination and there can be two possibilities for the sign of the parameter V_0. Each of them determine different dynamics in the evolution of the field equations. A positive sign leads to a period of inflation followed by one of deflation, whereas the opposite sign implies the contrary. In the first case, the values of V_0 we have found as viable to meet the COBE constrain, are not in agreement with those found in [39] by several orders of magnitude. The second case corresponding to vacuum domination has proved unrealistic to have a viable model of inflation since this process does not end and we do not have other mechanism to finish it as in the usual hybrid inflation scenario.

As mentioned before, the previous analysis was made under the assumption that the axion played no significant role during inflation. In section 5., the study of the evolution with the full dynamics is addressed, the results of this section indicate more possibilities to take into account for the analysis as a natural consequence of the inclusion of the axion. The only parameter of the potential that can dominate the evolution of the system is V_0, which leads to an eternal period of inflation and has, as a consequence, to be discarded as a viable model of inflation. This keeps a resemble of the result found previously in sections 4.6. and 4.7..

These results seem to be generic in superstrings theory, implying that if we would like to relate the moduli fields with the inflaton, dark energy or dark matter, the model could fit observations either during the inflationary epoch or during the structure formation, but the challenge is to derive a model which fit our observing universe during the whole history of the universe. Otherwise, superstrings theory have to give alternative candidates for these fields and explain why we do not see the moduli fields in our observations. Another possibility to explore is to consider a potential capable of stabilizing the σ_i fields in Eq (1) instead of considering them constants. In this case the dynamics should correspond to the standard cosmology but with more complicated field equations.

In the second part of this work we propose an alternative interpretation of the dilaton field in the type IIB supergravity on the $\mathbf{T}^6/\mathbb{Z}_2$ orientifold model with fluxes [29]. This alternative interpretation allowed us to compare this model with the ΛCDM one, which has been very successful in its predictions. The result is that, at lest in some models, radiation seems to be subdominant everywhere, provoking difficulties to explain big bang nucleosynthesis. Even when we see that in this particular toy model, the behavior seems to be generic for many strings theories. Nevertheless, we find a particular model where the radiation is dominant, but we have to fine tune the initial conditions. If this is the case for all string compactifications then it is possible that the dilaton and axion fields could not be able to be interpreted as dark matter or dark energy, thus ether we should seek other candidates and explain why we don't see the dilaton and axion scalar fields in our observations, or we have to explain big bang nucleosynthesis using the conditions given by superstrigs theory we showed here, or we have to look for a mechanism to eliminate the coupling between dilaton and axion with matter at very early times. This last option is maybe more realistic. Even if we solve the radiation dominance problem, there are some differences between ΛCDM and superstrings theory between $10^2 < z < 10^3$, because string theory predicts around 16 millon years of densities oscillations during the dark age. Nevertheless, both models are very similar at late times, between $0 < z < 10^2$, maybe the only difference during this last period is their predictions on substructure formation and galaxies centers. While CDM predicts much more substructure in the universe and very sharp density profiles, scalar fields predict few substructure and almost constant density profiles in centers of galaxies. The confirmation of this observations could decide between these two models.

We are aware that this is orientifold model is still a toy model and it would be interesting to study more realistic compactifications (including brane and orientifold configurations) and see if our results, including that of the dark matter density profiles, survive and become a general feature of string theory. If this is the case, this alternative interpretation of the fields of the theory might permit to establish a contact of string theory with the astrophysics phenomenology of dark matter, *i.e.*, its contact with future astrophysical and cosmological observations. We conclude that this interpretation can give us a closer understanding of superstrings theory with cosmology.

8. Acknowledgments

The authors wish to thank Luis Ureña-López for the deep reading of the paper, Norma Quiroz and Cesar Terrero for many helpful discussions and for pointing out to us the reference [29]. E.R. wants to thank Andrew Liddle and Anupam Mazumdar for their comments and revision of an earlier draft of this paper. Many thanks also to Andrew Frey for his comments on the previous version. The numerical computations were carried out in the "Laboratorio de Super-Cómputo Astrofísico (LaSumA) del Cinvestav" and the UNAM supercluster Kanbalam. This work was partly supported by CONACyT México, under grants 49865-F, and I0101/131/07 C-234/07, Instituto Avanzado de Cosmologia (IAC) collaboration.

References

[1] A. G. Riess *et al.*, *Astron.J.* **116** (1998)1009. Neta A. Bahcall, Jeremiah P. Ostriker, Saul Perlmutter and Paul J.Steinhardt, *Science* **284**, (1999)1481-1488. arXiv:astro-ph/9906463.

[2] B. P. Schmidt, *et al.*, Astrophys. J. **507**, 46 (1998); A. G. Riess, *et al.*, Astron. J. **116**, 1009 (1998); S. Perlmutter *et al.*, Astrophys. J. **517**, 565 (1999).

[3] Gerald Cleaver. Advances in Space Research, Volume 35, Issue 1, 2005, Pages 106-110

[4] Joel R. Primack. Proceedings of 5th International UCLA Symposium on Sources and Detection of Dark Matter, Marina del Rey, February 2002, ed. D. Cline

[5] M. B. Green, J. H. Schwarz and E. Witten, *Superstring Theory. Vol. 2: Loop Amplitudes, Anomalies And Phenomenology,* Cambridge University Press, Cambridge, 1987; J. Polchinski, *String Theory* Vol. 2, Cambridge University Press, Cambridge, 1998.

[6] Thibault Damour, Alexander Vilenkin. Phys. Rev. Lett. 78 (1997) 2288-2291

[7] Kiwoon Choi. Phys.Rev. D62 (2000) 043509 Kiwoon Choi. arXiv:hep-ph/9912218 Simeon Hellerman, Nemanja Kaloper, Leonard Susskind. JHEP 0106 (2001)003 arXiv:hep-th/0104180

[8] Y.M. Cho and Y.Y. Keum in Mod. Phys. Lett. A13, 109 (1998) and Class. Quant. Grav. 15, 907 (1998)

[9] Rainer Dick. arXiv:hep-th/9609190 Rainer Dick. Mod.Phys.Lett.A12:47-55,1997 P. J. E. Peebles and A. Vilenkin. Phys. Rev. D **60**, 103506 (1999) T. Matos and F. S. Guzman,"Scalar fields as dark matter in spiral galaxies," Class. Quant. Grav. **17**, L9 (2000) arXiv:gr-qc/9810028. Tirthabir Biswas, Robert Brandenberger, Anupam Mazumdar, Tuomas Multamaki. arXiv:hep-th/0507199

[10] K. Dasgupta, G. Rajesh and S. Sethi, "M Theory, Orientifolds and G-Flux", JHEP **9908** 023 (1999); G. Curio and A. Krause, " Four-Flux and Warped Heterotic M-Theory Compactifications", Nucl. Phys. B **602** 172 (2001),hep-th/0012152; S.B. Giddings, S. Kachru and J. Polchinski, "Hierarchies from Fluxes in String Compactifications", Phys. Rev. D **66** 106006 (2002); B.S. Acharya, "A Moduli Fixing Mechanism in M-theory", arXiv:hep-th/0212294; S. Kachru, M.B. Schulz and S. Trivedi, "Moduli Stabilization from Fluxes in Simple IIB Orientifold", JHEP **0310**, 007 (2003); A.R. Frey and J. Polchinski, "$N = 2$ Warped Compactifications", Phys. Rev. D **65** 126009 (2002); R. Blumenhagen, D. Lüst and T.R. Taylor, "Moduli Stabilization in Chiral Type IIB Orientifold Models with Fluxes", Nucl. Phys. B **663** 319 (2003); J.F.G. Cascales, M.P. García del Moral, F. Quevedo and A.M. Uranga, "Realistic D-brane Models on Warped Throats: Fluxes, Hierarchies and Moduli Stabilization", JHEP **0402** 031 (2004).

[11] R. Brandenberger, Y.-K. E. Cheung and S. Watson, "Moduli Stabilization with String Gases and Fluxes", arXiv:hep-th/0501032.

[12] K. Becker, M. Becker and A. Krause, "M-Theory Inflation from Multi M5-Brane Dynamics", Nucl. Phys. B **715** 349 (2005), hep-th/0501130.

[13] A.R. Frey, A. Mazumdar, "Three Form Induced Potentials, Dilaton Stabilization, and Running Moduli". Phys. Rev. D **67**:046006,2003. hep-th/0210254

[14] T. Matos and L. A. Urena-Lopez, "Quintessence and scalar dark matter in the universe," Class. Quant. Grav. **17**, L75 (2000) arXiv:astro-ph/0004332. V. Sahni and L. Wang, Phys. Rev. D **63**, (2000). A. Arbey, J. Lesgourgues and P. Salati, Phys. Rev. D **64**, 123528 (2001). *Ibid*, **65**, 083514 (2002).

[15] T. Matos and L. A. Ureña-López, Phys. Rev. D **63**, 063506 (2001).

[16] Jian-gang Hao, Xin-zhou Li. Phys.Rev. D67 (2003) 107303. L.P. Chimento, Ruth Lazkoz. Phys. Rev. Lett. 91 (2003) 211301. German Izquierdo, Diego Pavon. astro-ph/0505601. S. V. Sushkov.Phys.Rev. D71 (2005) 043520.

[17] E. J. Copeland, A. R. Liddle, and J. E. Lidsey, Phys. Rev. D **64**, 023509 (2001).

[18] A. R. Liddle and D. H. Lyth, *Cosmological Inflation and Large Scale Structure* (Cambridge University Press, Cambridge, 2000).

[19] J. E. Lidsey, T. Matos and L. A. Urena-Lopez,"The inflaton field as self-interacting dark matter," Phys. Rev. D **66**, 023514 (2002) arXiv:astro-ph/0111292.

[20] Erandy Ramirez, Tonatiuh Matos and Israel Quiros. "Inflation from IIB Superstrings with Fluxes." Int. J. Mod. Phys. A24, (2009), 2431-2452 arXiv:0708.3644.

[21] M. S. Turner Phys. Rev. D **28**, 1243 (1983)

[22] Tonatiuh Matos and José-Rubén Luévano. In preparation.

[23] T. Matos and L. A. Urena-Lopez,"A further analysis of a cosmological model of quintessence and scalar dark matter," Phys. Rev. D **63**, 063506 (2001) arXiv:astro-ph/0006024.

[24] F. S. Guzmán and L. A. Ureña-López, Phys. Rev. D **68**, 024023 (2003). arXivs:astro-ph/0303440. Guzmán, F.S., & Ureña-López, L.A. 2004, Phys. Rev. **D 69**, 124033.

[25] W. Hu, R. Barkana and A. Gruzinov, Phys. Rev. Lett. **85**, 1158 (2000).

[26] A. Bernal, T. Matos and D. Nuñez, "Flat Central Density Profiles from Scalar Field Dark Matter Halo," arXiv:astro-ph/0303455.

[27] M. Alcubierre, F. S. Guzman, T. Matos, D. Nunez, L. A. Urena-Lopez and P. Wiederhold,"Galactic collapse of scalar field dark matter," Class. Quant. Grav. **19**, 5017 (2002) arXiv:gr-qc/0110102.

[28] Navarro, J., Frenk, C. S., and White, S. S. M., 1997, ApJ, 490, 493 W. J. G. de Blok, S. S. MacGaugh, A. Bosma and V. C. Rubin, ApJ **552**, L23 (2001). S. S. MacGaugh, V. C. Rubin and W. J. G. de Blok, AJ **122**, 2381 (2201). W. J. G. de Blok, S. S. MacGaugh and V. C. Rubin, AJ **122**, 2396 (2001). Salucci, P., Walter, F., and Borriello, A., 2002. arXilv:astro-ph/0206304 Borriello, Salucci, Danese 2002, MNRAS 341, 1109 G. Gentile, A. Burkert, P. Salucci, U. Klein, F. Walter. arXiv:astro-ph/0506538 Monaco, L, Bellazzini, M., F.R. Ferraro F.R., Pancino, E. Mon.Not.Roy.Astron.Soc. 356 (2005) 1396-1402. Spekkens Kristine, Giovanelli Riccardo, Haynes Martha P. arXiv:astro-ph/0502166

[29] A.R. Frey, A. Mazumdar, "Three Form Induced Potentials, Dilaton Stabilization, and Running Moduli". Phys. Rev. D **67**:046006,2003. arXiv:hep-th/0210254.

[30] B. R. Greene, K. Schalm, G. Shiu. Nucl. Phys. **B584** (2000), 480-508, arXiv:hep-th/0004103.

[31] S. B. Giddings, S. Kachru, J. Polchinski. Phys. Rev. **D66** (2002), 106006, arXiv:hep-th/0105097.

[32] A. R. Frey, J. Polchinski. Phys. Rev. **D65** (2002) 126009, arXiv:hep-th/0201029.

[33] K. Becker, M. Becker. Nucl. Phys. **B477** (1996) 155-167,arXiv:hep-th/9605053.

[34] H. Verlinde. Nucl. Phys. **B580** (2000), 264-274, arXiv:hep-th/9906182.

[35] C. S. Chan, P. L. Paul, H. Verlinde, Nucl. Phys. **B581** (2000), 156-164, arXiv:hep-th/0003236.

[36] S. B. Giddungs, S. Kachru, J. Polchinski, Phys. Rev. **D66** (2002) 106006, arXiv:hep-th/0105097.

[37] R. Allahverdi, J. García-Bellido, K. Enqvist, A. Mazumdar, Phys. Rev. Lett. **97**, 191304 (2006).

[38] R. Allahverdi, K. Enqvist, J. García-Bellido, A. Jokinen, A. Mazumdar, JCAP **0706**, **019**, 2007.

[39] T. Matos, J.-R. Luévano, H. García-Compeán, A. Vázquez. Int. J. Mod.Phys. A23:1949-1962,2008,
arXiv:hep-th/0511098.

[40] A. Linde, Phys. Rev. D **62**, 103511, (1993). arXiv:astro-ph 9307002.

[41] E. J. Copeland, A. R. Liddle, D. H. Lyth, E. D. Stewart, D. Wands, Phys. Rev. D **49**, 12, 6410 (1994). arXiv:astro-ph:9401011.

[42] L. E. Mendes, A. R. Liddle, Phys. Rev. D **62**, 103511-1 (2000). arXiv:astro-ph/0006020.

[43] A. R. Liddle, P. Parsons, J. D. Barrow, Phys. Rev. D**50**, 7222 (1994). arXiv:astro-ph/9408015.

[44] L. Randall, R. Sundrum, Phys. Rev. Lett. **83**, 3370 (1999).

[45] L. Randall, R. Sundrum, Phys. Rev. Lett. **83**, 4690 (1999).

[46] T. Matos, J.-R. Luévano, L. A. Ureña-López, "Dynamics of a scalar field dark matter with a cosh potential".

[47] R. Maartens, D. Wands, B. A. Bassett, I. P. C. Heard., Phys. Rev. D **62**, 041301 (2000).

[48] A. R. Liddle, D. H. Lyth, Phys. Rep. **231**,1 (1993).

[49] F. Lucchin, S. Matarrese, Phys. Rev. D **32**, 1316 (1985).

[50] J. M. Cline, C. Grojean, G. Servant, Phys. Rev. Lett. **83**, 4245 (1999).

[51] A. R. Liddle, D. Parkinson, S. Leach, P. Mukherjee, Phys. Rev. D **74**, 083512 (2006). arXiv:astro-ph/0607275.

[52] B. Spokoiny, Phys. Lett.**B315**, 40-45, (1993).

In: Superstring Theory in the 21st Century...
Editor: Gerold B. Charney, pp.125-133
ISBN 978-1-61668-385-6

Chapter 4

BOSONIC STRING AND STRING FIELD THEORY: A SOLUTION USING THE HOLOMORPHIC REPRESENTATION*

C.G.Bollini and M.C.Rocca
Departamento de Física, Fac. de Ciencias Exactas,
Universidad Nacional de La Plata,
C.C. 67 (1900) La Plata. Argentina

Abstract

In this paper we show that the holomorphic representation is appropriate for description in a consistent way string and string field theories, when the considered number of component fields of the string field is finite. A new Lagrangian for the closed string is obtained and shown to be equivalent to Nambu-Goto's Lagrangian. We give the notion of anti-string, evaluate the propagator for the string field, and calculate the convolution of two of them.

PACS: 03.65.-w, 03.65.Bz, 03.65.Ca, 03.65.Db.

1. Introduction

In a series of papers [1, 2, 3, 4, 5] we have shown that Ultradistribution theory of Sebastiao e Silva [6, 7, 8] permits a significant advance in the treatment of quantum field theory. In particular, with the use of the convolution of Ultradistributions we have shown that it is possible to define a general product of distributions (a product in a ring with divisors of zero) that sheds new light on the question of the divergences in Quantum Field Theory. Furthermore, Ultradistributions of Exponential Type (UET) are adequate to describe Gamow States and exponentially increasing fields in Quantun Field Theory [9, 10, 11].

In four recent papers ([12, 13, 14, 15]) we have demonstrated that Ultradistributions of Exponential type provide an adequate framework for a consistent treatment of string and string field theories. In particular, a general state of the closed string is represented by UET

*This work was partially supported by Consejo Nacional de Investigaciones Cient'ıficas and Comisión de Investigaciones Científicas de la Pcia. de Buenos Aires; Argentina.

of compact support, and as a consequence the string field is a linear combination of UET of compact support (CUET). Thus a sting field theory result be a superposition of infinitely many of fields. The corresponding development is convergent due to that the superposition of infinitely many of complex Dirac's deltas and its derivatives is convergent.

However for experimental purposes is suitable consider the string field theory as a superposition of a finite but sufficient great number n of fields.

The resultant theory can be described in a simplified way with the use of the holomorphic representation [20]. In this case we can not consider a superposition of infinitely many of fields due to that we can not assure the convergence of the corresponding development in powers of the variable z.

In this paper we show that holomorphic representation provides an adequate method for a consistent simplified treatment of closed bosonic string. In particular, a general state of the closed bosonic string is represented by an polynomial function of a given number of complex variables.

This paper is organized as follows: In section 2 we give a new Lagrangian for bosonic string and solve the corresponding Euler-Lagrange's equations for closed bosonic string. In section 3 we give a new representation for the states of the string using the holomorphic representation. In section 4 we give expressions for the field of the string, the string field propagator and the creation and annihilation operators of a string and a anti-string. In section 5, we give expressions for the non-local action of a free string and a non-local interaction lagrangian for the string field similar to $\lambda\phi^4$ in Quantum Field Theory. Also we show how to evaluate the convolution of two string field propagators. In section 6 we realize a discussion of the principal results.

2. The Constraints for a Bosonic String

As is known the Nambu-Goto Lagrangian for the bosonic string is given by ([17])

$$\mathcal{L}_{NG} = T\sqrt{(\dot{X}\cdot X')^2 - \dot{X}^2 X'^2} \tag{2.1}$$

where

$$\begin{cases} X_\mu = X_\mu(\tau,\sigma)\ ;\ \dot{X}_\mu = \partial_\tau X_\mu\ ;\ X'_\mu = \partial_\sigma X_\mu \\ X_\mu(\tau,0) = X_\mu(\tau,\pi) \\ -\infty < \tau < \infty\ ;\ \ 0 \leq \sigma \leq \pi \end{cases} \tag{2.2}$$

If we use the constraint

$$(\dot{X} - X')^2 = 0 \tag{2.3}$$

we obtain:

$$\dot{X}^4 + X'^4 = 4(\dot{X}\cdot X')^2 - 2\dot{X}^2 X'^2 \geq 0 \tag{2.4}$$

On the other hand

$$(\dot{X}^2 - X'^2)^2 = \dot{X}^4 + X'^4 - 2\dot{X}^2 X'^2 \tag{2.5}$$

and from (2.4) we have

$$4\mathcal{L}_{BS}^2 = T^2(\dot{X}^2 - X'^2)^2 = 4T^2[(\dot{X}\cdot X')^2 - \dot{X}^2 X'^2] = 4\mathcal{L}_{NG}^2 \geq 0 \tag{2.6}$$

As a consequence of (2.6):

$$\mathcal{L}_{NG} = T\sqrt{(\dot{X}\cdot X')^2 - \dot{X}^2X'^2} = \frac{T}{2}|\dot{X}^2 - X'^2| = \mathcal{L}_{BS} \tag{2.7}$$

We then see that is sufficient to use only one constraint to obtain the Lagrangian for a bosonic string theory from the Nambú-Goto Lagrangian. Another constraint from which (2.6) follows is

$$(\dot{X} + X')^2 = 0 \tag{2.8}$$

Thus, the problem for the bosonic string reduces to:

$$\begin{cases} \mathcal{L} = \frac{T}{2}|\dot{X}^2 - X'^2| \\ (\dot{X} + X')^2 = 0 \\ X_\mu(\tau, 0) = X_\mu(\tau, \pi) \end{cases} \tag{2.9}$$

or

$$\begin{cases} \mathcal{L} = \frac{T}{2}|\dot{X}^2 - X'^2| \\ (\dot{X} - X')^2 = 0 \\ X_\mu(\tau, 0) = X_\mu(\tau, \pi) \end{cases} \tag{2.10}$$

The Euler-Lagrange equations for (2.9) and (2.10) are respectively:

$$4\delta(\dot{X}^2 - X'^2)[(\dot{X}\cdot\ddot{X} - X'\cdot\dot{X}')\dot{X}_\mu - (X'\cdot\dot{X}' - X'\cdot X'')X'_\mu]+$$

$$Sgn(\dot{X}^2 - X'^2)(\ddot{X} - X'') + \lambda(\ddot{X} + 2\dot{X}' + X'') = 0 \tag{2.11}$$

$$4\delta(\dot{X}^2 - X'^2)[(\dot{X}\cdot\ddot{X} - X'\cdot\dot{X}')\dot{X}_\mu - (X'\cdot\dot{X}' - X'\cdot X'')X'_\mu]+$$

$$Sgn(\dot{X}^2 - X'^2)(\ddot{X} - X'') + \lambda(\ddot{X} - 2\dot{X}' + X'') = 0 \tag{2.12}$$

where λ is a Lagrange multiplier.
Let X_μ be given by:

$$X_\mu = Sgn(\dot{Y}^2 - Y'^2)Y_\mu \tag{2.13}$$

where

$$\begin{cases} Y_\mu(\tau,\sigma) = y_\mu + l^2 p_\mu \tau + \frac{il}{2}\sum\limits_{n=-\infty\,;\,n\neq 0}^{\infty} \frac{a_n}{n} e^{-2in(\tau-\sigma)} \\ p^2 = 0 \end{cases} \tag{2.14}$$

or

$$\begin{cases} Y_\mu(\tau,\sigma) = y_\mu + l^2 p_\mu \tau + \frac{il}{2}\sum\limits_{n=-\infty\,;\,n\neq 0}^{\infty} \frac{\tilde{a}_n}{n} e^{-2in(\tau+\sigma)} \\ p^2 = 0 \end{cases} \tag{2.15}$$

(2.14) satisfy

$$\dot{Y}_\mu + Y'_\mu = p_\mu \tag{2.16}$$

and (2.15)

$$\dot{Y}_\mu - Y'_\mu = p_\mu \tag{2.17}$$

For both we have:

$$\dot{X}^2 - X'^2 = \dot{Y}^2 - Y'^2 \neq 0 \tag{2.18}$$

and then

$$(\dot{X}^2 - X'^2)^2 = (\dot{Y}^2 - Y'^2)^2 \neq 0 \tag{2.19}$$

((2.13), (2.14)) and ((2.13), (2.15)) are solutions of (2.11) and (2.12) respectively. To prove this we take into account that for ((2.13), (2.14)) we have

$$\ddot{X}_\mu = -\dot{X}' = X'' \tag{2.20}$$

and for ((2.13), (2.15))

$$\ddot{X}_\mu = \dot{X}' = X'' \tag{2.21}$$

To quantum level we have respectively for (2.14) and (2.15):

$$\begin{cases} Y_\mu(\tau,\sigma) = y_\mu + l^2 p_\mu \tau + \frac{il}{2} \sum\limits_{n=-\infty\,;\,n\neq 0}^{\infty} \frac{a_{n\mu}}{n} e^{-2in(\tau-\sigma)} \\ p^2|\phi >= 0 \end{cases} \tag{2.22}$$

and

$$\begin{cases} Y_\mu(\tau,\sigma) = y_\mu + l^2 p_\mu \tau + \frac{il}{2} \sum\limits_{n=-\infty\,;\,n\neq 0}^{\infty} \frac{\tilde{a}_{n\mu}}{n} e^{-2in(\tau+\sigma)} \\ p^2|\phi >= 0 \end{cases} \tag{2.23}$$

where $|\Phi>$ is the physical state of string.
In terms of creation and annihilation operators we have for (2.22) and (2.23):

$$\begin{cases} Y_\mu(\tau,\sigma) = y_\mu + l^2 p_\mu \tau + \frac{il}{2} \sum\limits_{n>0} \frac{b_{n\mu}}{\sqrt{n}} e^{-2in(\tau-\sigma)} - \frac{b^+_{n\mu}}{\sqrt{n}} e^{2in(\tau-\sigma)} \\ p^2|\phi >= 0 \end{cases} \tag{2.24}$$

$$\begin{cases} Y_\mu(\tau,\sigma) = y_\mu + l^2 p_\mu \tau + \frac{il}{2} \sum\limits_{n>0} \frac{\tilde{b}_{n\mu}}{\sqrt{n}} e^{-2in(\tau+\sigma)} - \frac{\tilde{b}^+_{n\mu}}{\sqrt{n}} e^{-2in(\tau+\sigma)} \\ p^2|\phi >= 0 \end{cases} \tag{2.25}$$

where:

$$[b_{\mu m}, b^+_{\nu n}] = \eta_{\mu\nu}\delta_{mn} \tag{2.26}$$

$$[\tilde{b}_{\mu m}, \tilde{b}^+_{\nu n}] = \eta_{\mu\nu}\delta_{mn} \tag{2.27}$$

A general state of the string can be written as:

$$\begin{aligned} |\phi >= [a_0(p) + a^{i_1}_{\mu_1}(p) b^{+\mu_1}_{i_1} + a^{i_1 i_2}_{\mu_1\mu_2}(p) b^{+\mu_1}_{i_1} b^{+\mu_2}_{i_2} + \ldots + \ldots \\ + a^{i_1 i_2 \ldots i_n}_{\mu_1\mu_2\ldots\mu_n}(p) b^{+\mu_1}_{i_1} b^{+\mu_2}_{1_2} \ldots b^{+\mu_n}_{i_n} + \ldots + \ldots]|0> \end{aligned} \tag{2.28}$$

or

$$\begin{aligned} |\phi >= [a_0(p) + a^{i_1}_{\mu_1}(p) \tilde{b}^{+\mu_1}_{i_1} + a^{i_1 i_2}_{\mu_1\mu_2}(p) \tilde{b}^{+\mu_1}_{i_1} \tilde{b}^{+\mu_2}_{i_2} + \ldots + \ldots \\ + a^{i_1 i_2 \ldots i_n}_{\mu_1\mu_2\ldots\mu_n}(p) \tilde{b}^{+\mu_1}_{i_1} \tilde{b}^{+\mu_2}_{1_2} \ldots \tilde{b}^{+\mu_n}_{i_n} + \ldots + \ldots]|0> \end{aligned} \tag{2.29}$$

where:

$$p^2 a^{i_1 i_2 \ldots i_n}_{\mu_1 \mu_2 \ldots \mu_n}(p) = 0 \tag{2.30}$$

It is immediate to prove that ((2.13), (2.14)) and ((2.13), (2.15)) are solutions of Nambu-Goto equations on physical states. (Nambu-Goto equations arise from Euler-Lagrange equations corresponding to the Lagrangian (2.1), and it is easy to prove that the currently used solution for the closed string movement is not solution of Nambu-Goto equations due to the fact that Virasoro operators L_n and $\tilde{L}_n$ does not annihilate the physical states for $n < 0$ and moreover, does not form a set of commuting operators).

3. A Representation of the States of the Closed String

Let $A(\mathbb{R}^2)$ be the complex Euclidean space defined as (see ref.[19] for a definition of complex Euclidean space)

$$A(\mathbb{R}^2) = \left\{ f(z)/f(z) \text{ is analytic } \wedge \frac{i}{2}\int f(z)\overline{f(z)}e^{-z\overline{z}}dz\, d\overline{z} < \infty \right\} \tag{3.1}$$

supplied with the complex scalar product (see ref.[20]):

$$< f(z), g(z) >= \frac{i}{2}\int f(z)\overline{g(z)}e^{-z\overline{z}}dz\, d\overline{z} = \tag{3.2}$$

Taking into account that:

$$\frac{i}{2}dz \wedge d\overline{z} = \frac{i}{2}(dx + idy) \wedge (dx - idy) =$$

$$\frac{1}{2}dx \wedge dy - dy \wedge dx = dx \wedge dy \tag{3.3}$$

where $\wedge$ denotes the outer product of two 1-forms, we have for the scalar product (3.2)

$$< f(z), g(z) >= \iint_{-\infty}^{\infty} f(z)\overline{g(z)}e^{-z\overline{z}}dx\, dy = \int_{-\infty}^{\infty}\int_{0}^{2\pi} f(z)\overline{g(z)}\rho e^{-\rho^2}d\theta\, d\rho \tag{3.4}$$

Let $Z(\mathbb{R}^2)$ be the complex Euclidean space defined as:

$$Z(\mathbb{R}^2) = \left\{ f(z)/z^p \frac{d^q f(z)}{d^q z} \in A(\mathbb{R}^2) \right\} \tag{3.5}$$

supplied with the scalar product (3.2), where p and q are natural numbers. For $f \in Z(\mathbb{R}^2)$ we have:

$$< \frac{df(z)}{dz}, g(z) >= \frac{i}{2}\int \frac{df(z)}{dz}\overline{g(z)}e^{-z\overline{z}}dz\, d\overline{z} =$$

$$\frac{i}{2}\int f(z)\overline{zg(z)}e^{-z\overline{z}}dz\, d\overline{z} =< f(z), zg(z) > \tag{3.6}$$

If we define:

$$a = \frac{d}{dz} \quad ; \quad a^+ = z \tag{3.7}$$

we obtain:

$$[a, a^+] = 1 \tag{3.8}$$

Representation (3.7) is called the holomorphic representation (see ref.[20]) for annihilation and creation operators.

The vacuum state annihilated by d/dz is the number $1/\sqrt{\pi}$ and the orthonormalized states obtained by successive application z to $1/\sqrt{\pi}$ are:

$$F_n(z) = \frac{z^n}{\sqrt{\pi\ n!}} \tag{3.9}$$

Using this representation a general state of the string can be written as:

$$\phi(x, \{z\}) = a_0(x) + a^{i_1}_{\mu_1}(x) z^{\mu_1}_{i_1} + a^{i_1 i_2}_{\mu_1 \mu_2}(x) z^{\mu_1}_{i_1} z^{\mu_2}_{i_2} + ... + ...$$

$$+ a^{i_1 i_2 ... i_n}_{\mu_1 \mu_2 ... \mu_n}(x) z^{\mu_1}_{i_1} z^{\mu_2}_{1_2} ... z^{\mu_n}_{i_n} \tag{3.10}$$

where $\{z\}$ denotes $(z_{1\mu}, z_{2\mu}, ..., z_{n\mu})$.
The functions $a^{i_1 i_2 ... i_n}_{\mu_1 \mu_2 ... \mu_n}(x)$ are solutions of

$$\Box a^{i_1 i_2 ... i_n}_{\mu_1 \mu_2 ... \mu_n}(x) = 0 \tag{3.11}$$

4. The String Field

According to (2.25), (2.25) and section 3 the equation for the string field is given by:

$$\Box \Phi(x, \{z\}) = (\partial_0^2 - \partial_1^2 - \partial_2^2 - \partial_3^2) \Phi(x, \{z\}) = 0 \tag{4.1}$$

where $\{z\}$ denotes $(z_{1\mu}, z_{2\mu}, ..., z_{n\mu}, ..., ...)$, and Φ is a analytic function in the set of variables $\{z\}$. Thus we have:

$$\Phi(x, \{z\}) = [A_0(x) + A^{i_1}_{\mu_1}(x) z^{\mu_1}_{i_1} + A^{i_1 i_2}_{\mu_1 \mu_2}(x) z^{\mu_1}_{i_1} z^{\mu_2}_{i_2} + ... + ...$$

$$+ A^{i_1 i_2 ... i_n}_{\mu_1 \mu_2 ... \mu_n}(x) z^{\mu_1}_{i_1} z^{\mu_2}_{1_2} ... z^{\mu_n}_{i_n} \tag{4.2}$$

where the quantum fields $A^{i_1 i_2 ... i_n}_{\mu_1 \mu_2 ... \mu_n}(x)$ are solutions of

$$\Box A^{i_1 i_2 ... i_n}_{\mu_1 \mu_2 ... \mu_n}(x) = 0 \tag{4.3}$$

The propagator of the string field can be expressed in terms of the propagators of the component fields:

$$\Delta(x - x', \{z\}, \{\overline{z}'\}) = \Delta_0(x - x') + \Delta^{i_1 j_1}_{\mu_1 \mu_2}(x - x') z^{\mu_1}_{i_1} \overline{z}'^{\nu_1}_{j_1} + ... + ...+$$

$$\Delta^{i_1 ... i_n j_1 ... j_n}_{\mu_1 ... \mu_n \nu_1 ... \nu_n}(x - x') z^{\mu_1}_{i_1} ... z^{\mu_n}_{i_n} \overline{z}'^{\nu_1}_{j_1} ... \overline{z}'^{\nu_n}_{j_n} \tag{4.4}$$

For the fields $A^{i_1 i_2 ... i_n}_{\mu_1 \mu_2 ... \mu_n}(x)$ we have:

$$A^{i_1 i_2 ... i_n}_{\mu_1 \mu_2 ... \mu_n}(x) = \int_{-\infty}^{\infty} a^{i_1 i_2 ... i_n}_{\mu_1 \mu_2 ... \mu_n}(k) e^{-ik_\mu x^\mu} + b^{+ i_1 i_2 ... i_n}_{\mu_1 \mu_2 ... \mu_n}(k) e^{ik_\mu x^\mu}\ d^3k \tag{4.5}$$

We define the operators of annihilation and creation of a string as:

$$a(k,\{z\}) = a_0(k) + a^{i_1}_{\mu_1}(k)z^{\mu_1}_{i_1} + ... + ...+$$

$$a^{i_1...i_n}_{\mu_1...\mu_n}(k)z^{\mu_1}_{i_1}...z^{\mu_n}_{i_n} \tag{4.6}$$

$$a^+(k^{'},\{\overline{z}^{'}\}) = a^+_0(k^{'}) + a^{+j_1}_{\nu_1}(k^{'})\overline{z}^{'\nu_1}_{j_1} + ... + ...+$$

$$a^{+j_1...j_n}_{\nu_1...\nu_n}(k^{'})\overline{z}^{'\nu_1}_{j_1}...\overline{z}^{'\nu_n}_{j_n} \tag{4.7}$$

and the annihilation and creation operators for the anti-string

$$b(k,\{\overline{z}\}) = b_0(k) + b^{i_1}_{\mu_1}(k)\overline{z}^{\mu_1}_{i_1} + ... + ...+$$

$$b^{i_1...i_n}_{\mu_1...\mu_n}(k)\overline{z}^{\mu_1}_{i_1}...\overline{z}^{\mu_n}_{i_n} \tag{4.8}$$

$$b^+(k^{'},\{z^{'}\}) = b^+_0(k^{'}) + b^{+j_1}_{\nu_1}(k^{'})z^{'\nu_1}_{j_1} + ... + ...+$$

$$b^{+j_1...j_n}_{\nu_1...\nu_n}(k^{'})z^{'\nu_1}_{j_1}...z^{'\nu_n}_{j_n} \tag{4.9}$$

If we define

$$[a^{i_1...i_n}_{\mu_1...\mu_n}(k), a^{+j_1...j_n}_{\nu_1..\nu_n}(k^{'})] = f^{i_1...i_nj_1...j_n}_{\mu_1...\mu_n\nu_1...\nu_n}(k)\delta(k-k^{'}) \tag{4.10}$$

the commutations relations are

$$[a(k,\{z\}), a^+(k^{'},\{\overline{z}^{'}\})] = [f_0(k) + f^{i_1j_1}_{\mu_1\nu_1}(k)z^{\mu_1}_{i_1}\overline{z}^{'\nu_1}_{j_1} + ... + ...$$

$$f^{i_1...i_nj_1...j_n}_{\mu_1...\mu_n\nu_1...\nu_n}(k)z^{\mu_1}_{i_1}...z^{\mu_n}_{i_n}\overline{z}^{'\nu_1}_{j_1}...\overline{z}^{'\nu_n}_{j_n}]\delta(k-k^{'}) \tag{4.11}$$

and for the anti-string:

$$[b^{i_1...i_n}_{\mu_1...\mu_n}(k), b^{+j_1...j_n}_{\nu_1..\nu_n}(k^{'})] = g^{i_1...i_nj_1...j_n}_{\mu_1...\mu_n\nu_1...\nu_n}(k)\delta(k-k^{'}) \tag{4.12}$$

the commutations relations are

$$[b(k,\{\overline{z}\}), b^+(k^{'},\{z^{'}\})] = [g_0(k) + g^{i_1j_1}_{\mu_1\nu_1}(k)\overline{z}^{\mu_1}_{i_1}z^{'\nu_1}_{j_1} + ... + ...$$

$$g^{i_1...i_nj_1...j_n}_{\mu_1...\mu_n\nu_1...\nu_n}(k)\overline{z}^{\mu_1}_{i_1}...\overline{z}^{\mu_n}_{i_n}z^{'\nu_1}_{j_1}...z^{'\nu_n}_{j_n}]\delta(k-k') \tag{4.13}$$

With this anihilation and creation operators we can write:

$$\Phi(x,\{z\}) = \int_{-\infty}^{\infty} a(k,\{z\})e^{-ik_\mu x^\mu} + b^+(k\{z\})e^{ik_\mu x^\mu}\, d^3k \tag{4.14}$$

5. The Action for the String Field

The action for the free bosonic bradyonic closed string field is:

$$S_{free} = \frac{i^n}{2^n} \int \int_{-\infty}^{\infty} \partial_\mu \Phi(x, \{z\}) e^{-\{z\}\cdot\{\overline{z}\}} \partial^\mu \Phi^+(x, \{z\})\, d^3x\, \{dz\}\, \{d\overline{z}\} \tag{5.1}$$

A possible interaction is given by:

$$S_{int} = \lambda \frac{i^n}{2^n} \int \int_{-\infty}^{\infty} \Phi(x, \{z\}) e^{-\{z\}\cdot\{\overline{z}\}} \Phi^+(x, \{z\}) e^{-\{\overline{z}\}\cdot\{z\}} \Phi(x, \{z\}) \times$$

$$e^{-\{z\}\cdot\{\overline{z}\}} \Phi^+(x, \{z\})\, d^3x\, \{dz\}\, \{d\overline{z}\} \tag{5.2}$$

Both, S_{free} and S_{int} are non-local as expected.

The convolution of two propagators of the string field is:

$$\hat{\Delta}(k, \{z_1\}, \{\overline{z}_2\}) * \hat{\Delta}(k, \{z_3\}, \{\overline{z}_4\}) \tag{5.3}$$

where $*$ denotes the convolution of Ultradistributions of Exponential Type on the k variable only. With the use of the result

$$\frac{1}{\rho} * \frac{1}{\rho} = -\pi^2 \ln \rho \tag{5.4}$$

($\rho = x_0^2 + x_1^2 + x_2^2 + x_3^2$ in euclidean space)

and

$$\frac{1}{\rho \pm i0} * \frac{1}{\rho \pm i0} = \mp i\pi^2 \ln(\rho \pm i0) \tag{5.5}$$

($\rho = x_0^2 - x_1^2 - x_2^2 - x_3^2$ in minkowskian space)

the convolution of two string field propagators is finite.

6. Discussion

We have shown that holomorphic representation is appropriate for the description in a consistent way string and string field theories. By means of a new Lagrangian for the closed string strictly equivalent to Nambu-Goto Lagrangian we have obtained a movement equation for the field of the string and solve it. We shown that this string field is a polynomial in the variables z. We evaluate the propagator for the string field, and calculate the convolution of two of them, taking into account that string field theory is a non-local theory. For practical calculations and experimental results we have given expressions that involve only a finite number of variables.

As a final remark we would like to point out that our formulas for convolutions follow from general definitions. They are not regularized expressions

References

[1] D. G. Barci, G. Bollini, L. E. Oxman, M. C. Rocca. (1998). Int. J. of Theor. Phys. **37**, N.12, 3015.

[2] C. G. Bollini, T. Escobar and M. C. Rocca. (1999). Int. J. of Theor. Phys. **38**, 2315.

[3] C. G Bollini and M.C. Rocca. (2004). Int. J. of Theor. Phys. **43**, 59.

[4] C. G. Bollini and M. C. Rocca. (2004). Int. J. of Theor. Phys. **43**, 1019.

[5] C. G. Bollini and M. C. Rocca: (2007). Int. J. of Theor. Phys. **46**, 3030.

[6] J. Sebastiao e Silva. (1958). Math. Ann. **136**, 38.

[7] M. Hasumi. (1961) Tôhoku Math. J. **13**, 94.

[8] R. F. Hoskins and J. Sousa Pinto. (1994). "Distributions, Ultradistributions and other Generalised Functions". Ellis Horwood.

[9] C. G. Bollini, L. E. Oxman, M. C. Rocca. (1994). J. of Math. Phys. **35**, N. 9, 4429.

[10] C. G. Bollini, O. Civitarese, A. L. De Paoli, M. C. Rocca. (1996). J. of Math. Phys. **37**, 4235.

[11] A. L. De Paoli, M. Estevez, H. Vucetich, M. C. Rocca. (2001). Infinite Dimensional Analysis, Quantum Probability and Related Topics **4**, N.4, 511.

[12] C. G Bollini and M. C. Rocca. (2008). Int. J. of Theor. Phys. **47**, 1409.

[13] C. G. Bollini and M. C. Rocca. (2009). Int. J. of Theor. Phys. **48**, 1053.

[14] C. G. Bollini, A. L. De Paoli and M. C. Rocca. (2008) "World Sheet Superstring and Superstring Field Theory: a new solution using Ultradistributions of Exponential Type". La Plata preprint.

[15] C. G. Bollini and M. C. Rocca. (2009) "A Solution to Non-Linear Equations of Nambu-Goto String".La Plata preprint.

[16] I. M. Gel'fand and N. Ya. Vilenkin. (1964) "Generalized Functions" **Vol. 4**. Academic Press.

[17] Y. Nambu. (1970).in Symmetries and quark models, ed. R. Chand. Gordon and Breach. (1970). Lectures at the Copenhagen Symposium.

[18] T. Goto. (1971). Prog. Theor. Phys. **46**, 1560 (1971).

[19] A. N. Kolmogorov and S. V. Fomin. (1975). "Elementos de la Teoria de Funciones y del Analisis Funcional". Translated from the Russian version. Mir, Moscu.

[20] L. D. Faddeev and A. A. Slavnov. (1980). "Gauge Fields". The Benjamin/Cummings Publishing Company, Inc..

In: Superstring Theory in the 21st Century...
Editor: Gerold B. Charney, pp.135-160
ISBN 978-1-61668-385-6

Chapter 5

THERMAL AND QUANTUM INDUCED EARLY SUPERSTRING COSMOLOGY

F. Bourliot[1,*] ***J. Estes***[1,†] ***C. Kounnas***[2‡] ***and H. Partouche***[1§]
[1]Centre de Physique Théorique, Ecole Polytechnique,
F–91128 Palaiseau cedex, France
[2] Laboratoire de Physique Théorique, Ecole Normale Supérieure,
24 rue Lhomond, F–75231 Paris cedex 05, France

Abstract

In this work, we review the results of Refs [1] –[5] dedicated to the description of the early Universe cosmology induced by quantum and thermal effects in superstring theories. The present evolution of the Universe is described very accurately by the standard Λ-CDM scenario, while very little is known about the early cosmological eras. String theory provides a consistent microscopic theory to account for such missing epochs. In our framework, the Universe is a torus filled with a gas of superstrings. We first show how to describe the thermodynamical properties of this system, namely energy density and pressure, by introducing temperature and supersymmetry breaking effects at a fundamental level by appropriate boundary conditions.

We focus on the intermediate period of the history: After the very early "Hagedorn era" and before the late electroweak phase transition. We determine the back-reaction of the gas of strings on the initially static space-time, which then yields the induced cosmology. The consistency of our approach is guaranteed by checking the quasi-staticness of the evolution. It turns out that for arbitrary initial boundary conditions at the exit of the Hagedorn era, the quasi-static evolutions are universally attracted to radiation-dominated solutions. It is shown that at these attractor points, the temperature, the inverse scale factor of the Universe and the supersymmetry breaking scale evolve proportionally. There are two important effects which result from the underlying string description. First, initially small internal dimensions can be spontaneously decompactified during the attraction to a radiation dominated Universe. Second, the radii of internal dimensions can be stabilized.

*E-mail address:Francois.Bourliot@cpht.polytechnique.fr
†E-mail address: John.Estes@cpht.polytechnique.fr
‡E-mail address: Costas.Kounnas@lpt.ens.fr
§E-mail address: Herve.Partouche@cpht.polytechnique.fr

1. Why and How Studying Superstring Cosmology?

We are aware of the existence of four interactions: Gravity, weak interaction, electromagnetism and strong interaction. Though gravity is the oldest known one, it is still the less well understood. In quantum field theory, electromagnetism and the weak interaction have successfully been embedded into the electroweak interaction [6], while the strong interaction is very well described by QCD. A common wisdom is that at very high energies, these three interactions combine into a Grand Unified Theory (GUT) based on a gauge group $G_{\rm GUT}$ containing at least the standard model group $SU(3)_{\rm strong} \times SU(2)_{\rm weak} \times U(1)_{\rm elec}$. As one lowers the energy/temperature, the interactions unified in $G_{\rm GUT}$ start to separate and the standard model emerges. In particular, the electroweak phase transition takes place at low energy and implies that electromagnetism and the weak interaction split. While theoretically promising, there is still no experimental evidence for the existence of such a GUT.

[†] Unité mixte du CNRS et de l'Ecole Polytechnique, UMR 7644.

[‡] Unité mixte du CNRS et de l'Ecole Normale Supérieure associée à l'Université Pierre et Marie Curie (Paris 6), UMR 8549.

Though interesting for studying the last three interactions, the above GUT scenario has an important drawback: It does not include gravity. The oldest known interaction does not merge with the others in such a consistent quantum field theory. However, it is not currently possible to access experimentally domains of energies relevant for testing a quantum theory of gravity, except maybe through astrophysical and cosmological observations of phenomena involving very high energies. Consequently, the realm of quantum gravity has been, and is still, the one for theorists. Over the last thirty years, there has been various attempts to formulate a consistent quantum theory of gravity. Not all of them try to also embed gravity with the other three forces in a Quantum Theory Of Everything (QTOE). String Theory [7] is a candidate for a QTOE, and it is an important open problem to realize within it not only the standard model of particle physics but also the known cosmology of our Universe.

What we mean by "known cosmology" can be described by the Cosmological Standard Model, dubbed the Λ-CDM model (Cold Dark Matter). The latter proposes an history of our space-time and predicts its ultimate fate. Starting just after its birth, the Universe underwent a very fast period of acceleration, called *Inflation*. The latter diluted the primordial inhomogeneities, topological relics and rendered the space flat. At the end of inflation, the Universe undergoes a short period of "reheating" and an era of domination by radiation then appears. As the temperature lowers, symmetry breaking phase transitions occur (and in particular the electroweak breaking) and the fundamental particles acquire a mass via the Higgs mechanism. As the Universe cools, hadrons such as the proton and neutron start to form. Progressively, matter appears as the results of thermonuclear reactions. Consequently, after some time, matter dominates and leads to structure formation. However, it no longer dominates today since there are observational evidences that our Universe is slightly accelerating [8]. This is implemented theoretically speaking, by introducing a tiny cosmological constant. For a wider discussion on standard cosmology and the history of our Universe, see [9].

General Relativity is the theory which best describes gravity in a classical setting. Supposing the Universe to be 4-dimensional with coordinates x^μ ($\mu = 0, \ldots, 3$), the Einstein equation takes the form

$$G_{\mu\nu} := R_{\mu\nu} - \frac{1}{2} R\, g_{\mu\nu} = T_{\mu\nu}, \tag{1.1}$$

in appropriate units such that $c = 1$ and $8\pi G = 1$. In (1.1), $R_{\mu\nu}$ is the Ricci tensor, while $T_{\mu\nu}$ is the stress-energy tensor. Due to inflation, the Λ-CDM model treats the Universe as homogeneous and isotropic at sufficiently large scales, and spatially flat. It is then possible to show [10] that the metric describing such a space-time is given by the following FLRW (Friedmann-Lemaître-Robertson-Walker) one:

$$ds^2 = -N(t)^2 dt^2 + a(t)^2 \sum_{\mu=1}^{3} (dx^\mu)^2, \tag{1.2}$$

where $a(t)$ is the *scale factor*. In a homogeneous and isotropic space-time, the stress-energy tensor takes, in the perfect fluid approximation, the form

$$T_{\mu\nu} = (P + \rho)\, u_\mu u_\nu + P g_{\mu\nu}, \tag{1.3}$$

where u^μ is the 4-speed of the cosmic fluid, satisfying $u^\mu u_\mu = -1$. ρ is the energy density and P the pressure. The Λ-CDM model describes phenomenologically the observed features of our Universe by coupling Einstein gravity to a matter sector whose field content, potentials and kinetic terms are constrained only by observations. These sources are described by perfect fluids of species $i = 1, 2, \ldots$ characterized by their densities ρ_i and pressures P_i related by equations of state $P_i = \omega_i\, \rho_i$, with parameter ω_i. To be concrete, the Λ-CDM model states that 97% of the energetic content of the Universe is described by dark matter (27%) and a cosmological constant Λ (70%) for dark vacuum energy. While there is much indirect experimental evidence for them, these two quantities still lack direct measurement so that their exact nature is still unknown. Theoretically speaking, there is a wide diversity of scenarios with both dark matter and cosmological constant candidates (see [11, 12] for a cosmological scenario trying to explain both dark matter and cosmological constant). The remaining 3% of the content of the Universe is spanned between pressureless non-relativistic baryonic matter and radiation, the second being a tiny fraction of the first in our present epoch. Though very interesting, the phenomenological approach of the Λ-CDM model lacks an underlying microscopic derivation.

It is a challenge for String Theory to provide such a foundation. However, despite considerable efforts toward unraveling string cosmology over the last few years, still very little is known about the dynamics of strings in time-dependent settings. Indeed, it seems difficult to obtain time-dependent solutions in string theory at the classical level. After extensive studies in the framework of superstring compactifications, the obtained results appear to be unsuitable for cosmology. In most cases, the classical ground states correspond to static Anti-de Sitter like or flat backgrounds but not time-dependent ones. The same situation appears to be true in the effective supergravity theories. Naively, the results obtained in this direction may yield to the conclusion that cosmological backgrounds are unlikely to be found in superstring theory. However, quantum and thermal corrections are neglected in the classical string/supergravity regime. Actually, it turns out that in certain cases, the quantum and thermal corrections are under control [2]-[5] at the full string level and that cosmological evolutions at finite temperature can be generated dynamically at the quantum level. The purpose of the present work is to review them and show how some of the weak points of phenomenological approaches can be explored and analyzed concretely in a consistent theoretical framework. In particular, we will describe how one can find the energy density and pressure from microscopic arguments, by studying the canonical ensemble of a gas of strings.

In order to understand how cosmological solutions arise naturally in this context, we first consider classical supersymmetric flat backgrounds in 4 dimensions. They are obtained from the 10-dimensional space-time in which superstrings are living by compactifying 6 directions. The study of the thermodynamics of a gas of superstring states filling this background makes sense at the quantum level only (this is well known from Planck, when he introduced the notion of quanta to solve the UV catastrophe problem in black body physics). At finite temperature, the quantum and thermal fluctuations produce a non-zero free energy density which is computable perturbatively at the full string level. Note that in this context, it can be determined order by order in the Riemann surface genus expansion without the UV ambiguities encountered in the analogous computation in quantum field theory. An energy density and pressure can be derived from the free energy. Their back-reaction on the space-time metric and moduli fields (the continuous parameters of the models) gives rise to specific cosmological evolutions. In this review, the above strategy is restricted to the domain of temperatures lower than the Hagedorn temperature and higher than the electroweak breaking scale to be specified in the next paragraphs. In this intermediate regime, the evolution of the Universe is found to converge to a radiation dominated era.

More interesting models are those where space-time supersymmetry is spontaneously broken at a scale M before finite temperature is switched on. With the supersymmetry breaking mechanism we consider, the stringy quantum corrections are under control in a way similar to the thermal ones [2] –[5]. In large classes of models, the back-reaction of the quantum and thermal corrections on the space-time metric and the moduli fields induces a cosmological evolution which is attracted to a radiation dominated era. The latter is characterized by a temperature and a supersymmetry breaking scale that evolve proportionally to the inverse of the scale factor, $T(t) \propto M(t) \propto 1/a(t)$.

In the context of string theory, we can study much higher energies than the ones tested so far. We are then limited by the appearance of a *Hagedorn phase transition* at ultra high temperature [13]. It is a consequence of the exponential growth of the number of states that can be thermalized at high temperature and implies a divergence of the canonical thermal partition function. The latter can be computed in Matsubara formalism *i.e.* in Euclidean time compactified on an circle of circumference β, the inverse temperature. The Hagedorn instability is signaled by string modes wrapping the Euclidean time circle that become tachyonic (*i.e.* with negative (mass)^2) when the temperature is above the Hagedorn temperature T_H [14, 15]. The possible existence of an Hagedorn era in the very early Universe provides a possible alternative or at least a complementary point of view to inflation, as developed in [16]. However, we do not discuss this very high temperature regime here, and we will consider the physics at low enough temperatures compared to T_H to avoid the occurrence of a Hagedorn phase transition [17]. Note that models free of Hagedorn instabilities [18, 19] and still under computational control can also be constructed. In the present review, we bypass the Hagedorn era ambiguities by assuming that 3 large spatial directions have emerged before t_E (the exit time of the Hagedorn era), along with internal space directions whose size characterizes the scale of spontaneous supersymmetry breaking. Within this assumption, we parameterize our ignorance of the detailed physics in the Hagedorn era by considering arbitrary initial boundary conditions (IBC) for the fields at t_E.

At late cosmological times, when the temperature of the Universe is low enough, it is possible for an additional scale Q to become relevant. Q is the infrared renormalisation group invariant transmutation scale induced at the quantum level by the radiative corrections of the soft supersymmetry breaking terms at low energies [20]. When $T(t) \sim Q$, the electroweak phase transition takes place, $SU(2) \times U(1) \to U(1)_{elec}$. This starts to be the case at a time t_W and, for $t > t_W$, the supersymmetry breaking scale M is stabilized at a value close to Q. In earlier cosmological times where $M(t) \sim T(t) > Q$, the transmutation scale Q is irrelevant and the Universe is in the radiation era. It turns out that the electroweak symmetry breaking transition is very sensitive to the specifics of the string background considered, while in the earlier radiation era the results are fairly robust. We restrict our analysis to the intermediate cosmological times:

$$t_E \ll t \ll t_W, \tag{1.4}$$

namely, after the exit of the Hagedorn era and before the electroweak symmetry breaking.

In section 2, we present the basics of our approach and apply it to the simplest examples where supersymmetry is spontaneously broken by temperature effects only. In this class of models, the moduli (radii) of the internal space are held fixed, close to the string scale. In section 3, we analyze models where supersymmetry is spontaneously broken even at zero temperature. We take into account the dynamics of the supersymmetry breaking scale M which is a field and keep frozen the other moduli. This is only in section 4 that we show the latter hypothesis is consistent by taking into account the dynamics of internal radii that are not participating in the breaking of supersymmetry. The last section is devoted to our conclusions.

2. Basics of our Approach

2.1. Thermodynamics and Variational Principle

Let us first describe how thermodynamical results can be derived from general relativity. As a simple example, we consider the gas of a single bosonic state at temperature T in a 3-dimensional torus T^3, which is nothing but a box with periodic boundary conditions and large volume $V_{\text{box}} = (2\pi R_{\text{box}})^3$. The number of particles is not fixed and the canonical ensemble partition function Z_{th} is defined in terms of the Hamiltonian H and inverse temperature β. In second quantized formalism, Z_{th} can be expressed as a path integral

$$Z_{th} := \text{Tr}\, e^{-\beta H} = \int \mathcal{D}\varphi\, e^{-S_E[\varphi]}, \tag{2.5}$$

where S_E is the Euclidean action of the quantum field φ and the Euclidean time is compact with period β. The boundary condition along the Euclidean circle of the bosonic field φ is periodic (while a fermionic field would be anti-periodic). The thermal partition function can be written in terms of an infinite sum of connected or disconnected Feynmann graphs. Supposing the gas to be (almost) perfect, the particles do not interact (much) with themselves and we can approximate the result at one loop. The free energy takes the form

$$F = -\frac{\ln Z_{\rm th}}{\beta} \simeq -\frac{Z_{1-\rm loop}}{\beta}, \tag{2.6}$$

where $Z_{1-\rm loop}$ is the unique connected graph at 1-loop, the *bubble diagram*, a single propagator whose two ends are identified. Then, one can derive from F the energy density and pressure using standard thermodynamics identities.

However, it is also possible to use the fact that $Z_{1-\rm loop}$ is, from a quantum field theory point of view, the 1-loop vacuum-to-vacuum amplitude *i.e.* vacuum energy inside the box. Viewing the whole space we are living in as the box itself, the classical Einstein action must be corrected at 1-loop by a contribution to the "cosmological constant",

$$S = \int d^4x\sqrt{-g}\left(\frac{R}{2} + \frac{Z_{1-\rm loop}}{\beta V_{\rm box}}\right). \tag{2.7}$$

Note that in the above action, we are back to real time *i.e. Lorentzian* signature, by analytic continuation on the time variable. The stress-energy tensor is found by varying with respect to the metric,

$$T_{\mu\nu} = -\frac{2}{\sqrt{-g}}\frac{\delta}{\delta g^{\mu\nu}}\left(\sqrt{-g}\,\frac{Z_{1-\rm loop}}{\beta V_{\rm box}}\right). \tag{2.8}$$

Originally, the classical background is homogeneous and locally flat Minkowski space. Its metric is of the form (1.2) with laps function $N = \beta$ and scale factor $a = 2\pi R_{\rm box}$, as follows from the analytic continuation of the Euclidean background in which the 1-loop vacuum-to-vacuum amplitude has been computed. With $Z_{1-\rm loop}$ a function of β and $V_{\rm box}$, the stress-energy tensor takes the form $T^\mu{}_\nu = {\rm diag}(-\rho, P, P, P)^\mu{}_\nu$, where

$$P = \frac{1}{\beta}\frac{\partial Z_{1-\rm loop}}{\partial V_{\rm box}} \equiv -\left(\frac{\partial F}{\partial V_{\rm box}}\right)_\beta \tag{2.9}$$

$$\rho = -\frac{1}{V_{\rm box}}\frac{\partial Z_{1-\rm loop}}{\partial \beta} \equiv \frac{1}{V_{\rm box}}\left(\frac{\partial(\beta F)}{\partial\beta}\right)_{V_{\rm box}}. \tag{2.10}$$

In these expressions, the right hand sides follow from Eq. (2.6) and reproduce the standard thermodynamical results. When $Z_{1-\rm loop}$ is proportional to $V_{\rm box}$ *i.e.* the free energy is extensive, these relations simplify to

$$P = -\mathcal{F} \qquad \rho = T\frac{\partial P}{\partial T} - P, \tag{2.11}$$

where $\mathcal{F} := \dfrac{F}{V_{\rm box}}$ is the free energy density and the relation between ρ and P is the state equation.

The above approach has the advantage to allow to go farther than deducing the thermodynamical identities. As long as the 1-loop sources ρ and P are small perturbations of the classically homogeneous and isotropic static background, one can find their back-reaction on the space-time metric by solving the Einstein equation (1.1). In other words, a quasi-static evolution $\beta(t), a(t)$ is found. This hypothesis amounts to supposing that the evolution is a sequence of thermodynamical equilibria *i.e.* that it is slow enough for the temperature to be remain homogenous.

We could have been satisfied by this quantum field theory approach in general relativity if an important difficulty would not arise: In most cases, $Z_{1-\text{loop}}$ is actually divergent ! This is always the case for a single bosonic (or fermionic) field. However, suppose the gas contains two species of free particles, one bosonic of mass M_B and one fermionic of mass M_F. The 1-loop vacuum-to-vacuum amplitude is found to be

$$Z_{1-\text{loop}} = \beta V_{\text{box}} \int_0^{+\infty} \frac{dl}{2l} \frac{1}{(2\pi l)^2} \sum_{\tilde{m}_0} \left(e^{-\frac{l}{2}M_B^2} - (-)^{\tilde{m}_0} e^{-\frac{l}{2}M_F^2} \right) e^{-\frac{\beta^2 \tilde{m}_0^2}{2l}}, \qquad (2.12)$$

where l is a "Schwinger parameter", the proper time of each particle when it runs into a loop wrapped $\tilde{m}_0$ along the Euclidean time circle. If $M_B \neq M_F$ (that could arise from a spontaneous breaking of supersymmetry), the contribution to this integral with $\tilde{m}_0 = 0$ is divergent in the UV, $l \to 0$. We conclude that only the exactly supersymmetric spectrum $M_B = M_F$ gives a well defined free energy.

As is well know, the amplitudes in string theory are free of UV divergences and one can expect that the above approach applied in this context will give a perfectly well established framework to describe thermodynamics in time-dependent backgrounds. Actually, we are going to see that string theory provides a rigorous microscopic derivation of the sources ρ and P and their equation of state.

2.2. Supersymmetric String Models at Finite *T*

We start to implement the ideas sketched at the end of the previous section on simple string theory models in 4-dimensional Minkowski space-time, where supersymmetry is spontaneously broken by thermal effects only. The case of models where supersymmetry is spontaneously broken even at zero temperature is addressed in the next section.

To be specific, we consider heterotic models but type II or type I ones can be treated similarly. To analyze the canonical ensemble of a gas of heterotic strings, we consider 10-dimensional backgrounds of the form

$$\underbrace{S^1(R_0)}_{\text{Euclidean time}} \times \underbrace{T^3(R_{\text{box}})}_{\text{space}} \times \underbrace{\mathcal{M}_6}_{\text{internal space}}, \qquad (2.13)$$

where $S^1(R_0)$ is the Euclidean time circle of perimeter $\beta = 2\pi R_0$, $T^3(R_{\text{box}})$ denotes the spatial part which is taken to be flat and compact *i.e.* a very large torus, and $\mathcal{M}_6$ represents the remaining internal manifold.

Classically, the vacuum energy vanishes. This is clear from the fact that the genus-0 vacuum-to-vacuum string amplitude is computed on the Riemann sphere, which is simply connected and thus cannot wrap the Euclidean time. Consequently, it cannot probe the temperature effects that are responsible of the breaking of supersymmetry. However, the genus-1 Riemann surface which is nothing but a torus has two cycles that can wrap the Euclidean time so that a non-trivial 1-loop contribution to the vacuum-to-vacuum energy arises, $Z_{1-\text{loop}}$. To be specific, we focus on the simplest example where $\mathcal{M}_6 = T^6$, which means that at $T = 0$ the model is $\mathcal{N} = 4$ supersymmetric in 4 dimensions for the heterotic case. Using world-sheet techniques, $Z_{1-\text{loop}}$ is given by [2]-[5]

$$Z_{1-\text{loop}} = \frac{\beta V_{\text{box}}}{(2\pi)^4} \int_{\mathcal{F}} \frac{d\tau d\bar{\tau}}{4(\text{Im}\,\tau)^3} \frac{1}{2} \sum_{a,b} (-1)^{a+b+ab} \frac{\vartheta^4[^a_b]}{\eta^4} \frac{\Gamma_{(6,22)}}{\eta^8 \bar{\eta}^{24}}$$

$$\sum_{\tilde{m}_0, n_0} e^{-\frac{\pi R_0^2}{\text{Im}\,\tau} |\tilde{m}_0 + n_0 \tau|^2} (-1)^{\tilde{m}_0 a + \tilde{n}_0 b + \tilde{m}_0 n_0}. \tag{2.14}$$

A few words might be helpful to understand this expression. On the 2-dimensional world-sheet of the heterotic string, there are left moving superstring waves and right moving bosonic ones. There are thus 10 left moving world-sheet fermions and bosons, and 26 right moving world-sheet bosons. However, ghosts cancel the contributions of two bosons (left and right) and two left fermions. The remaining fermions contribute the factor $\vartheta^4[^a_b](\tau)/\eta^4(\tau)$ with an appropriate spin-statistic phase $(-1)^{a+b+ab}$ depending on the integers a, b modulo 2 associated to the boundary conditions of the fermions along the two circles of the world-sheet torus. The contributions of the left and right moving world-sheet bosons correspond to the $\Gamma_{(6,22)}(\tau, \bar{\tau})/\eta^8(\tau)/\bar{\eta}^{24}(\bar{\tau})$ factor. $\Gamma_{(6,22)}$ is a lattice that corresponds to the 0-modes of the 16 right moving bosons (without left partners) and the six internal left and right moving bosons that realize the coordinates of the internal space T^6. In field theory, the loop can wrap $\tilde{m}_0$ times around the Euclidean time. In string theory, the closed string itself can also be wrapped n_0 times around $S^1(R_0)$. This is why we have a double discrete sum on arbitrary integers $\tilde{m}_0, n_0$ and a generalized phase $(-1)^{\tilde{m}_0 a + \tilde{n}_0 b + \tilde{m}_0 n_0}$ that involves the winding number n_0, as compared to Eq. (2.12), where space-time bosons (fermions) correspond to $a = 0$ ($a = 1$).

Another important difference between the field and string theory amplitudes, Eqs (2.12) and (2.14), is that the integration over the Schwinger parameter from 0 to $+\infty$ is replaced by an integral over the *fundamental domain* of $SL(2, \mathbb{Z})$,

$$\mathcal{F} = \left\{ \tau \in \mathbb{C} \;\Big/\; |\text{Re}\,\tau| \leq \frac{1}{2},\; \text{Im}\,\tau > 0,\; |\tau| \geq 1 \right\}, \tag{2.15}$$

which does not contain the line $\text{Im}\,\tau = 0$. Since the "Schwinger parameter" or proper time along the string world-sheet torus is $\text{Im}\,\tau$, there is no risk of any UV divergence. This property of the amplitude is only due to the extended nature of the string that provides a natural cut-off in the UV. Contrarily to the field theory case, the UV finiteness of the amplitude is guaranteed in all models, even when the spectrum at zero temperature is not supersymmetric (see section 3). However, both in field and string theory, the amplitude is finite in the IR (l and $\text{Im}\,\tau \to +\infty$) as long as there is no tachyon in the spectrum (*i.e.* no particle with $(\text{mass})^2 < 0$). A careful analysis shows that string states winding around the Euclidean time circle become tachyonic when $1/R_H < R_0 < R_H$, where $R_H = \dfrac{1+\sqrt{2}}{2}$ in $\sqrt{\alpha'}$ units [14] –[17], the string length we have set to 1 in Eq. (2.14) and now on for notational simplicity. R_H determines the Hagedorn temperature at which a phase transition occurs. As announced in section 1, we restrict ourselves to the study of epochs in the history of the Universe that follow the Hagedorn era and thus consider only regimes where $R_0 \gg 1$.

In the amplitude (2.14) (or (2.12) in field theory), the dominant contributions arise from the lightest states. The pure Kaluza-Klein (KK) states associated to the Euclidean time circle have masses of order $1/R_0$. Strings with non-trivial winding number n_0 around $S^1(R_0)$ get a contribution to their mass proportional to the length of the circle *i.e.* R_0. Similarly, the KK and winding states associated to the internal space $\mathcal{M}_6$ have masses contributions of order $1/R_I$ and R_I, where R_I denotes some generic radius (modulus) characterizing the size of $\mathcal{M}_6$. Finally, each string that oscillates has a contribution of order 1 to its mass. For simplicity in this section, we suppose that all radii R_I are satisfying the constraint

$$\frac{1}{R_0} \ll R_I \ll R_0 \tag{2.16}$$

that will be justified in section 4.[1] It follows that the towers of pure KK states along the Euclidean time are much lighter than any other states in the spectrum. Given that, the partition function (2.14) can be written as

$$Z_{1-\text{loop}} = \beta V_{\text{box}} \frac{1}{(2\pi R_0)^4} n_T\, c_4 + \cdots \text{ where } c_4 = \frac{1}{\pi^2} \sum_{\tilde{m}_0} \frac{1}{|2\tilde{m}_0 + 1|^4} = \frac{\pi^2}{48} \tag{2.17}$$

and the dots stand for contributions of order $\mathcal{O}(e^{-2\pi R_0})$ for the oscillating states, $\mathcal{O}(e^{-2\pi R_0/R_I})$ and $\mathcal{O}(e^{-2\pi R_0 R_I})$ for the KK and the winding states of $\mathcal{M}_6$, and $\mathcal{O}(e^{-2\pi R_0^2})$ for winding states around $S^1(R_0)$. These terms are all exponentially suppressed, compared to the dominant contribution. In Eq. (2.17), n_T is the number of massless boson-fermion pairs in the supersymmetric model when the temperature is not switched on. The constant c_4 is a dressing that accounts for the full towers of KK states along $S^1(R_0)$.

Let us determine the back-reaction of the non-trivial 1-loop vacuum energy on the originally static background. At order one in string perturbation theory, the low energy effective action at finite temperature is

$$S = \int d^4x \sqrt{-g_{st}} \left[e^{-2\phi} \left(\frac{R_{st}}{2} + 2(\partial\phi)^2 \right) + \frac{Z_{1-\text{loop}}}{\beta V_{box}} \right], \tag{2.18}$$

where ϕ is the dilaton in four dimensions. Compared to the general relativity case Eq. (2.7), the string "coupling constant" is actually the field $e^{2\phi}$. The choice in the definition of the metric tensor that gives rise to the above mixing of the dilaton and the Ricci curvature is referred as the string frame metric. In (2.18), we have kept constant the over massless fields of the string spectrum since the 1-loop source does not involve them in the present case (see the next sections for more general models). Their kinetic terms are thus vanishing.

The action may be converted to a more convenient "Einstein frame" by rescaling the metric as $g_{st\mu\nu} = e^{2\phi}\, g_{\mu\nu}$,

$$S = \int d^4x \sqrt{-g} \left[\frac{R}{2} - (\partial\phi)^2 - \mathcal{F} \right], \tag{2.19}$$

where

$$\mathcal{F} = -T^4\, n_T\, c_4 \qquad\qquad T = \frac{1}{2\pi R_0\, e^{-\phi}}. \tag{2.20}$$

[1] It will be shown that for arbitrary I.B.C. at the exit of the Hagedorn era, R_I is dynamically attracted to the interval $1/R_0 < R_I < R_0$ and then converges to a constant.

$\mathcal{F}$ and T which contain a dilaton dressing are the free energy density and temperature when they are measured in Einstein frame. Supposing that the back-reaction of the thermal sources induce a quasi-static evolution of the homogeneous and isotropic background, dilaton and temperature, we consider an ansatz

$$\begin{aligned} &ds^2 = -N(x^0)^2(dx^0)^2 + a(x^0)^2\left[(dx^1)^2 + (dx^2)^2 + (dx^3)^2\right], \qquad \phi(x^0), \\ &\text{where} \quad N(x^0) \equiv 2\pi R_0\, e^{-\phi} \equiv \frac{1}{T(x^0)}, \qquad a(x^0) \equiv 2\pi R_{box}\, e^{-\phi}. \end{aligned} \tag{2.21}$$

The derivation of the stress-energy tensor reaches

$$\rho = 3P \qquad \text{where} \qquad P = T^4\, n_T\, c_4, \tag{2.22}$$

which is nothing but Stefan's law for radiation. This was expected since under the hypothesis (2.16), all non-zero masses are of order 1 which is also the scale of T_H. Since we consider temperatures far below the Hagedorn one, only the massless states can be thermalized. The massive ones remain "cold" *i.e.* decoupled from the thermal system.

The equations of motion are easily solved. In terms of cosmological time such that $N dx^0 = dt$, the velocity of the dilaton, $\dot{\phi} \propto 1/a^3$, goes to zero when the Universe expands. The evolution is thus attracted to the particular solution where ϕ is constant *i.e.* the cosmology of a Universe filled by a radiation fluid:

$$a(t) = \sqrt{t} \times a_0 T_0 (n_T c_4)^{1/4} = \frac{1}{T(t)} \times a_0 T_0\,, \qquad \phi = cst., \tag{2.23}$$

where a_0, T_0 are integration constants. We shall refer to such an attractor as a Radiation Dominated Solution (RDS).

Though we started with the huge machinery of string theory, we finally ended with standard results when the Universe is filled with radiation, after we consider the dominant contribution of massless modes. The reader could be skeptical about the need to require such heavy tools to describe such simple physics. However, we remind that our approach gives a rigorous microscopic derivation of these results that will be generalized in sections 3 and 4 to models whose free energies are UV divergent in field theory. In addition, we are going to see that when the dynamics of other scalar fields is taken into account, the underlying string theory provides a connection between naively disconnected theories as seen from a field theory point of view. This string theoretic effect marks the novelty of our approach.

3. Non-supersymmetric String Models at Finite *T*

The aim of the previous section was to present the basic ideas of a string theory framework able to provide a microscopic origin for source terms for the gravitational (and moduli) fields. We considered a model $\mathcal{N} = 4$ supersymmetric when temperature is not switched on. In order to recover a non-supersymmetric physics at very low temperature *i.e.* late time from a cosmological point of view, we need to consider models whose spectra are not supersymmetric, even at $T = 0$. From a phenomenological point of view, we are particularly interested in models with $\mathcal{N} = 1$ spontaneously broken supersymmetry.

To switch on finite temperature, we have introduced periodic or antiperiodic boundary conditions on the Euclidean time circle for the string states, depending on their fermionic number ($a = 0$ for bosons and $a = 1$ for fermions in Eq. (2.14)). In a similar way, a spontaneous breaking of supersymmetry can be generated by non-trivial boundary conditions on internal circles $S^1(R_i)$ ($i = 4, \dots, 3+n$), using R-symmetry charges $a + Q_i$. Both finite temperature and supersymmetry breaking implemented this way can be thought as a string theoretic generalizations of Scherk-Schwarz compactifications [21]. Physically speaking, this can be thought as introducing non-trivial background fluxes along the cycles. Two mass scales then appear and are *a priori* time-dependent: The temperature $T \propto \frac{1}{2\pi R_0}$ and the supersymmetry breaking scale $M \propto \frac{1}{2\pi(\prod_i R_i)^{1/n}}$. The initially degenerate mass levels of bosons and fermions split by amounts proportional to T and/or M. This mass splitting is the signal of supersymmetry breaking and gives rise to a non-trivial free energy density at 1-loop. Note that in the pure thermal case, each originally degenerate boson-fermion pair gives a positive contribution to $Z_{1-\text{loop}}$ since it is always the fermion that is getting a mass shift (their momenta along $S^1(R_0)$ are half integer). However, when introducing supersymmetry breaking, bosons can have non-trivial R-symmetry charges and acquire masses bigger than the ones for fermions. Consequently, negative contributions to the vacuum energy can arise. This will play an important role for RDS to exist. In the following, we first consider simple examples of models with $n = 1$ before sketching some cases with $n = 2$ to observe how ratios of radii (commonly referred as "complex structure moduli") can be stabilized.

3.1. Supersymmetry Breaking Involving *n* = 1 Internal Dimension

We want to study the canonical ensemble of a gas of heterotic strings, where supersymmetry is spontaneously broken by non-trivial boundary conditions along the internal direction 4. To be specific, we focus on two kinds of backgrounds (2.13), with internal space

$$\text{(I)}: \quad \mathcal{M}_6 = S^1(R_4) \times S^1 \times \frac{T^4}{\mathbb{Z}_2} \quad \text{or} \quad \text{(II)}: \quad \mathcal{M}_6 = \frac{S^1(R_4) \times T^3}{\mathbb{Z}_2} \times T^2. \quad (3.24)$$

$\mathbb{Z}_2$ acts as $x^I \to -x^I$, where $I = 6, 7, 8, 9$ in case (I) and $I = 4, 5, 6, 7$ in case (II), and breaks explicitly half of the supersymmetries. Thus, the above models have spectra where $\mathcal{N} = 2$ supersymmetry is spontaneously broken to zero by the internal flux around $S^1(R_4)$ and the temperature effects. Models with $\mathcal{N} = 1 \to 0$ can also be analyzed by considering $\mathcal{M}_6 = \dfrac{S^1(R_4) \times T^5}{\mathbb{Z}_2 \times \mathbb{Z}_2}$. They share similar cosmological properties with the models in case (II).

As in the pure thermal case of section 2, tachyonic instabilities arise when R_0 or R_4 approach R_H. In order for the canonical ensemble to be well defined, we will restrict our analysis to regimes where $R_0 \gg 1$ and $R_4 \gg 1$. We also suppose that all internal radii that are not participating in the spontaneous breaking of supersymmetry satisfy

$$\frac{1}{R_0} \ll R_I \ll R_0 \qquad \frac{1}{R_4} \ll R_I \ll R_4 \qquad (I \neq 4), \qquad (3.25)$$

and remind the reader that section 4 is devoted to the justification that this is consistent with the dynamics of the R_I's. Under these hypothesis, the 1-loop partition function of the pure thermal case (2.17) is generalized to

$$Z_{1-\text{loop}} = \beta V_{\text{box}} \left(\frac{1}{(2\pi R_0)^4}\, n_T\, \hat{f}_T^{(4)}(z) + \frac{1}{(2\pi R_4)^4}\, n_V\, e^{-z} \hat{f}_T^{(4)}(-z) + \frac{1}{(2\pi R_0)^4}\, n'_T\, c_4 + \cdots \right), \tag{3.26}$$

where

$$\hat{f}_T^{(4)}(z) = \frac{\Gamma(5/2)}{\pi^{5/2}} \sum_{\tilde{k}_0, \tilde{k}_4 \in \mathbb{Z}} \frac{e^{4z}}{[(2\tilde{k}_0+1)^2 e^{2z} + (2\tilde{k}_4)^2]^{5/2}}, \qquad e^z = \frac{R_0}{R_4}. \tag{3.27}$$

In Eq. (3.26), R_0 and R_4 being large, the modes which are KK excitations along both $S^1(R_0)$ and $S^1(R_4)$ give dominant contributions corresponding to the two first terms in the parenthesis. In other words, while R_i/R_I and R_iR_I ($i = 0, 4$ and $I \neq 4$) are very large and give exponentially suppressed contributions we neglect, it is not necessary the case for R_0/R_4. As a consequence, the dimension full factors $1/R_0^4$ and $1/R_4^4$ are dressed with non-trivial functions of the "complex structure" ratio $e^z = R_0/R_4$. In case (I), n_T is the number of massless states (before we switch on finite temperature). n_V is the number of massless bosons minus the number of massless fermions (before we switch on finite temperature) and depends on the choice of R-symmetry charge $a + Q_4$ used to break spontaneously supersymmetry. Both n_T and n_V contain contributions arising from the untwisted and twisted sectors of the $\mathbb{Z}_2$-orbifold, while $n'_T = 0$. In case (II), n_T and n_V are defined similarly, but contain contributions from the untwisted sector only. On the contrary, the mass spectrum in the $\mathbb{Z}_2$-twisted sector does not depend on R_4 and supersymmetry between the associated states is spontaneously broken by thermal effects only. n'_T is then the number of massless boson/fermion pairs in the twisted sector (before we switch on finite temperature). To summarize, we have

$$n_T > 0\,, \quad -1 \leq \frac{n_V}{n_T} \leq 1\,, \quad n'_T = 0 \ \text{ in case (I)}\,, \quad n'_T > 0 \ \text{ in case (II)}. \tag{3.28}$$

As before, $Z_{1-\text{loop}}$ backreacts on the classical 4-dimensional Lorentzian background, via the effective field theory action,

$$S = \int d^4x \sqrt{-g_{st}} \left[e^{-2\phi} \left(\frac{R_{st}}{2} + 2(\partial\phi)^2 + \frac{1}{2}(\partial \ln R_4)^2 \right) + \frac{Z_{1-\text{loop}}}{\beta V_{\text{box}}} \right], \tag{3.29}$$

where the scalar kinetic term of R_4 is included since there is a non-trivial source at 1-loop for it. Before writing the equations of motion, it is useful to redefine the scalar fields as,

$$\Phi := \sqrt{\frac{2}{3}}\,(\phi - \ln R_4)\,, \qquad \phi_\perp := \frac{1}{\sqrt{3}}\,(2\phi + \ln R_4)\,, \tag{3.30}$$

and switch from string to Einstein frame metric. The action becomes

$$S = \int d^4x \sqrt{-g} \left[\frac{R}{2} - \frac{1}{2}\left((\partial\phi)^2 + (\partial\phi_\perp)^2\right) - \mathcal{F} \right], \tag{3.31}$$

where $\mathcal{F}$ is the free energy density,

$$\mathcal{F} = -T^4 \left(n_T \, \hat{f}_T^{(4)}(z) + n_V \, e^{3z} \hat{f}_T^{(4)}(-z) + n'_T \, c_4 \right) := -T^4 \, p(z), \tag{3.32}$$

and z can be expressed in terms of the temperature and supersymmetry breaking scales as,

$$e^z = \frac{M}{T}, \qquad T = \frac{1}{2\pi \, R_0 e^{-\phi}}, \qquad M = \frac{1}{2\pi R_4 \, e^{-\phi}} \equiv \frac{e^{\sqrt{\frac{3}{2}}\Phi}}{2\pi}. \tag{3.33}$$

Assuming an homogeneous and isotropic Universe with a flat 3-dimensional subspace as in Eq. (2.21), the 1-loop components of the stress-tensor are found to be:

$$P = T^4 \, p(z) \qquad \rho = T^4 \left(3p(z) - p_z(z) \right) := T^4 \, r(z), \tag{3.34}$$

where p_z stands for a derivation with respect to z, $p_z = \frac{\partial}{\partial z} p$. The fields are only time-dependent and their equations of motion are:

$$3H^2 = \frac{1}{2}\dot{\Phi}^2 + \frac{1}{2}\dot{\phi}_\perp^2 + \rho, \tag{3.35}$$

$$\dot{\rho} + 3H(\rho + P) + \sqrt{\frac{3}{2}}\, \dot{\Phi}\,(3P - \rho) = 0, \tag{3.36}$$

$$\ddot{\Phi} + 3H\dot{\Phi} = \frac{\partial P}{\partial \Phi} \equiv \sqrt{\frac{3}{2}}\,(3P - \rho), \tag{3.37}$$

$$\ddot{\phi}_\perp + 3H\dot{\phi}_\perp = 0 \qquad \Longrightarrow \qquad \dot{\phi}_\perp = \sqrt{2}\,\frac{c_\perp}{a^3}, \tag{3.38}$$

where $c_\perp$ is an integration constant. Denoting $\overset{\circ}{f} \equiv \dfrac{df}{d\ln a}$, we can first obtain the relation $\overset{\circ}{z} = \sqrt{\frac{3}{2}}\,\overset{\circ}{\Phi} - \frac{\overset{\circ}{T}}{T}$. Differentiating this identity and using the equations of motion, we can substitute Eq. (3.37) with an equation for z of the form

$$h(z, \overset{\circ}{z}, \overset{\circ}{\phi}_\perp) \left(\mathcal{A}(z)\overset{\circ\circ}{z} + \mathcal{B}(z)\overset{\circ}{z}{}^2 \right) + \mathcal{C}(z)\overset{\circ}{z} + V_z(z) = 0, \tag{3.39}$$

where we introduce the notion of an effective potential V:

$$V_z(z) = r - 4p. \tag{3.40}$$

The explicit expressions for $h(z, \overset{\circ}{z}, \overset{\circ}{\phi}_\perp)$, $\mathcal{A}(z)$, $\mathcal{B}(z)$ and $\mathcal{C}(z)$ can be found in Ref.[4]. Eq. (3.39) admits a static solution $z \equiv z_c$, $\phi_\perp \equiv cst.$ when the potential $V(z)$ admits a critical point z_c. It happens that the shape of $V(z)$ depends drastically on the parameters n_V/n_T and n'_T:

- In case (I), $n'_T = 0$ and three behaviors can arise, as depicted on Figure 1:
  - Case (I*a*): For $\dfrac{n_V}{n_T} < -\dfrac{1}{15}$, $V(z)$ increases.
  - Case (I*b*): For $-\dfrac{1}{15} < \dfrac{n_V}{n_T} < 0$, $V(z)$ has a unique minimum z_c, and $p(z_c) > 0$.

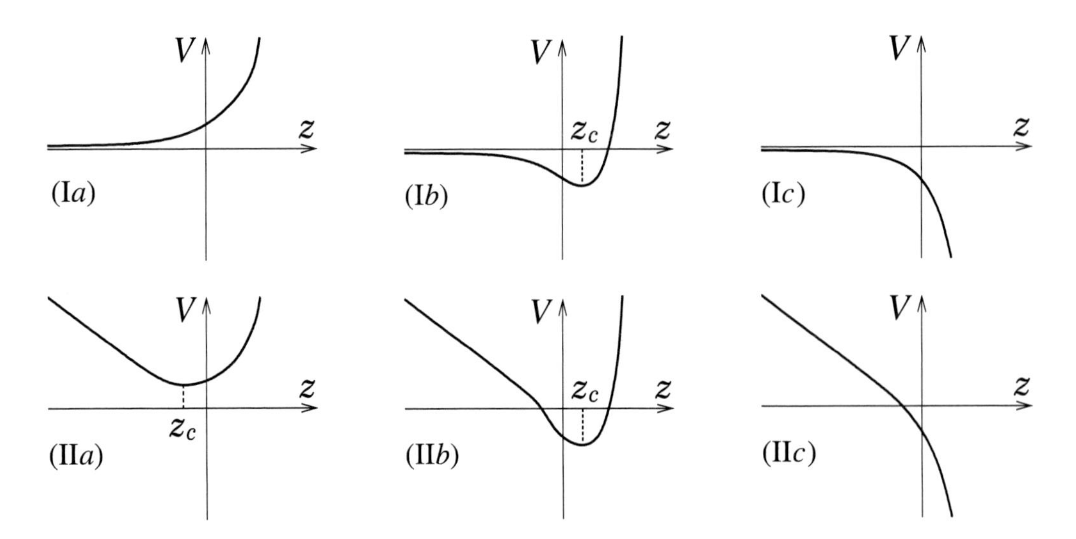

Figure 1: Different qualitative behaviors of $V(z)$. The cases (a), (b) and (c) correspond to $-1 < n_V/n_T < -1/15$, $-1/15 < n_V/n_T < 0$ and $0 < n_V/n_T < 1$. For models of type (I), an extremum exists in case (b) only. For models of type (II), an extremum occurs in cases (a) and (b) i.e. $n_V < 0$.

- Case (Ic): For $0 < \frac{n_V}{n_T}$, $V(z)$ decreases.

- In case (II), $n'_T > 0$. In the three above ranges, the behaviors differ from case (I) for large negative z, where the potentials are linearly decreasing (see Figure 1):
  - Case (IIa) & (IIb): For $n_V < 0$, $V(z)$ has a unique minimum z_c, and $p(z_c) > 0$.
  - Case (IIc): For $0 < n_V$, $V(z)$ decreases.

Thus, the cases (Ib), (IIb) and (IIa) admit a very particular solution where both scalars are constants *i.e.* $z \equiv z_c$ and $c_\perp = 0$. The conservation of the stress-tensor (3.36) and the Friedmann equation (3.35) gives then:

$$M(t) = T(t)\, e^{z_c} = \frac{1}{a(t)} \times a_0 M_0 \qquad \text{with} \qquad a(t) = \sqrt{\frac{t}{t_0}} \times a_0\,, \quad \phi_\perp = cst. \quad (3.41)$$

where a_0, M_0, t_0 are constants. This evolution is characterized by a temperature T, a spontaneous supersymmetry breaking scale M and an inverse scale factor that are proportional for all times, with $H^2 \propto 1/a^4$ and $\phi_\perp$ a modulus. It is thus an RDS.

Using the positivity properties of $h(z, \overset{\circ}{z}, \overset{\circ}{\phi}_\perp)$, $\mathcal{A}(z)$, $\mathcal{B}(z)$ and $\mathcal{C}(z)$, it is easy to check analytically that this RDS is stable for small perturbations, implying that it is a local attractor of the dynamics [4]. Global attraction is also true, but numerics are required to yield to this conclusion. Figure 2 gives an example of convergence to the RDS obtained for a generic choice of I.B.C..

In case (Ia), one can show analytically that when $z \ll -1$ and $|\overset{\circ}{z}| \ll 1$, the friction due to the expansion of the Universe implies $\overset{\circ}{z}$ to converge to 0. In other words, z is attracted

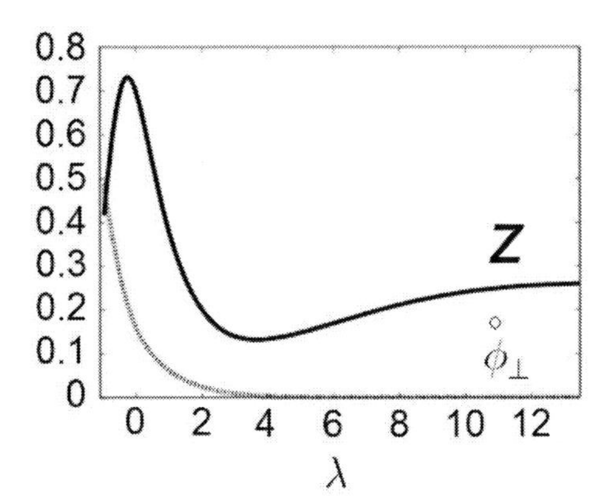

Figure 2: Example of damping oscillations of $z(\lambda)$ where $\lambda \equiv \ln a$ (solid curve) and convergence to zero of $\overset{\circ}{\phi}_\perp(\lambda)$ (dotted curve) illustrating the dynamical attraction towards the critical solution $z \equiv z_c \simeq 0.272$, $\dot{\phi}_\perp \equiv 0$. It correspond to $\frac{n_V}{n_T} = -0.02$, $n'_T = 0$ i.e. some case (Ib). The initial conditions are $(z_0, \overset{\circ}{z}_0, \overset{\circ}{\phi}_{\perp 0}) = (0.4, 0.8, 0.5)$.

to an arbitrary constant along the flat region of $V(z)$. The cosmological evolution is thus converging to the RDS (3.41), where $z_c \ll -1$ is a modulus whose value is now determined by the I.B.C.. A numerical study shows that for arbitrary I.B.C., z ends by sliding along its potential and enters the regime $z \ll -1$, $|\overset{\circ}{z}| \ll 1$ that yields to the above conclusions. It follows that the evolution is always attracted to the RDS. However, since $e^z = R_0/R_4 \ll 1$ with $R_0 \gg 1$, it is more natural to interpret the RDS from a 5-dimensional point of view *i.e.* to consider $S^1(R_4)$ as part of the space-time itself. In that case, R_4 does not appear in the definition of a scalar field M (or Φ), but is interpreted as a fifth component of the metric, $g_{st44} = (2\pi R_4)^2$. The attractor (3.41) is then rewritten as an RDS in 5 dimensions, where supersymmetry is spontaneously broken by thermal effects only: The scale factor $a'(t)$ of the directions 1, 2, 3, the scale factor $b(t)$ of the direction 4 and the temperature T' of the 5-dimensional Universe evolve proportionally for all times and one has $H'^2 \propto 1/a'^5$. The mechanism described in case (Ia) thus corresponds to the *dynamical decompactification* of an internal direction involved in the spontaneous breaking of supersymmetry.

Finally, in cases (Ic) and (IIc), one can show that the scale factor ends by decreasing (eventually after a turning point where $\dot{a} = 0$) and that ρ and P are formally diverging at late times [4, 5]. This implies that our underlying hypothesis of quasi-staticness of the evolution breaks down at some time, since the perturbations of the background are very large. Thermodynamics out of equilibrium should then be applied and is out of the scope of the present work.

As a conclusion, the originally static models based on internal backgrounds $\mathcal{M}_6$ given in Eq. (3.24) or of the form $\dfrac{S^1(R_4) \times T^5}{\mathbb{Z}_2 \times \mathbb{Z}_2}$ are giving rise to cosmological evolutions attracted to RDS if and only if the partition function $\mathbb{Z}_{1-\text{loop}}$ contains a negative contribution, $n_V \leq 0$.[2] In addition, the space-time dimension of the late time evolution is determined dynamically in case (I). It is important to mention that during the attraction to the RDS, there

[2] The limit case $n_V = 0$ can be seen to yield an RDS in 4 dimensions [4].

is no substantial period of accelerated expansion for the Universe *i.e.* this intermediate era cannot account for inflation. Moreover, it is easy to see numerically that the time needed to reach the RDS can exceed the age of our real Universe, especially in case (I). To avoid this problem, one can start with I.B.C. close enough to the radiation era or restrict to models in case (II), due to the replacement of the plateau for $z \ll -1$ by a steeper potential. Note that the more realistic models where $\mathcal{N}=1$ is spontaneously broken belong precisely to this class.

3.2. $n = 2$ Models and Complex Structure Stabilization

New phenomena can occur when the results of the previous section are extended to models with non-trivial boundary conditions along $n = 2$ internal directions, say 4 and 5. To be specific, let us analyze 4-dimensional Euclidean backgrounds with internal space

$$\mathcal{M}_6 = S^1(R_4) \times S^1(R_5) \times \mathcal{M}_4, \tag{3.42}$$

where $R_0, R_4, R_5 \gg 1$ to avoid any risk of Hagedorn-like phase transition. Again, we restrict ourselves to radii R_I of the space $\mathcal{M}_4$ satisfying

$$\frac{1}{R_i} \ll R_I \ll R_i \qquad (i = 0,4,5\,; \quad I \neq 4,5). \tag{3.43}$$

As before, the dominant contribution to the 1-loop partition function $Z_{1-\text{loop}}$ arises from the pure KK excitations along the circles $S^1(R_i)$ ($i = 0,4,5$), while the other modes give exponentially damped terms that can be safely neglected. By analogy with Eq. (3.27), $Z_{1-\text{loop}}$ can be expressed in terms of triple discrete sums involving two ratios of radii *i.e.* two "complex structures". A convenient choice for them is,

$$e^z := \frac{R_0}{\sqrt{R_4 R_5}}\,, \qquad e^Z := \frac{R_5}{R_4}\,. \tag{3.44}$$

In terms of the temperature T and the supersymmetry breaking scale M, the free energy density is found to take the following form,

$$\mathcal{F} = -T^4\, p(z,Z) \;\text{ where }\; e^z = \frac{M}{T}\,, \quad T = \frac{1}{2\pi\, R_0 e^{-\phi}}\,, \quad M = \frac{1}{2\pi\sqrt{R_4 R_5}\, e^{-\phi}}. \tag{3.45}$$

Depending on the precise way to break supersymmetry along the internal directions 4 and 5, different cosmological behaviors are found. In some cases one or the other of the directions 4 and 5 is spontaneously decompactified and we are back to a case already treated in the previous subsection (generalized in higher dimensions). In other cases, (z, Z) is found to converge to a critical point (z_c, Z_c) corresponding again to an RDS in 4 dimensions, where $T(t) \propto M(t) \propto 1/a(t)$. The new phenomenon encountered in such simple models is the dynamical stabilization of the complex structure $e^Z = R_5/R_4$.

4. Stabilization of Kähler Structures

In the previous sections, we have supposed that the internal radii-moduli R_I that are not participating in the spontaneous breaking of supersymmetry are bounded by the radii (and their inverses) that do participate in the breaking (see Eqs (2.16) and (3.25)). This hypothesis was fundamental to ague that the R_I's appear in the 1-loop free energy through exponentially suppressed terms only. Here, we would like to justify this assumption is consistent by analyzing the dynamics of the R_I's, the so-called "Kähler moduli", for arbitrary initial values and velocities.

Since R_I can *a priori* be large, the associated $S^1(R_I)$ may be treated as a space-time direction rather than an internal one. We therefore consider the framework of the previous sections with arbitrary number $d-1$ of large spatial directions and, for simplicity, we consider the dynamics of a single internal circle, $S^1(R_d)$. As an example, we introduce a spontaneous breaking of supersymmetry that involves $n=1$ internal circle, say in the direction 9, $S^1(R_9)$. Temperature is implemented as usually with non-trivial boundary conditions along the Euclidean time $S^1(R_0)$. Altogether, we consider the following 10-dimensional background in heterotic or type II superstring,

$$S^1(R_0)\times T^{d-1}(R_{\rm box})\times S^1(R_d)\times \mathcal{M}_{10-d-2}\times S^1(R_9), \tag{4.46}$$

where $R_0 \gg 1$ and $R_9 \gg 1$, while the remaining radii R_I of the internal manifold $\mathcal{M}_{10-d-2}$ satisfy

$$\frac{1}{R_0}\ll R_I \ll R_0 \qquad\qquad \frac{1}{R_9}\ll R_I\ll R_9 \qquad (I\neq d). \tag{4.47}$$

Since we do not consider a $\mathbb{Z}_2$ orbifold action on $S^1(R_9)$, we are actually considering models in case (I), in the notations of section 3 (see [5] for other cases).

Due to the T-duality on $S^1(R_d)$, $Z_{1-{\rm loop}}$ admits a symmetry $R_d\to 1/R_d$ that relates the regime $R_d\geq 1$ to $R_d\leq 1$. When $R_d\gg 1$, the light states are the KK modes along the directions $S^1(R_0)$, $S^1(R_9)$ and $S^1(R_d)$. As in section 3.2., their contribution to $Z_{1-{\rm loop}}$ involves two complex structures, say R_0/R_9 and R_9/R_d. On the contrary, when R_d approaches 1 from above, not only the winding but also the KK modes along $S^1(R_4)$ cease to contribute substantially, since their mass is 1 (in $\sqrt{\alpha'}$ units), up to a numerical factor. However, in heterotic string theory, this numerical factor happens to be 0 for the particular modes whose winding and momentum numbers are $\tilde{m}_d=n_d=\pm 1$ (the extended gauge symmetry point $U(1)\to SU(2)$ at $R_d=1$). In other words, while these specific states are super massive for generic values of R_d, their KK towers along $S^1(R_0)$, $S^1(R_9)$ do contribute when $R_d\simeq 1$. Defining

$$e^z:=\frac{R_0}{R_9}\,,\qquad e^\eta:=R_9\,,\qquad e^\zeta:=R_d, \tag{4.48}$$

and dropping terms that are exponentially suppressed in any regime of R_d, the string partition is found to be

$$Z_{1-{\rm loop}}=\beta V_{\rm box}\,\frac{1}{(2\pi R_0)^d}\,p(z,\eta,\zeta), \tag{4.49}$$

where p can be expressed in one way or another as,

$$p(z,\eta,\zeta) = n_T\left[\hat{f}_T^{(d)}(z) + k_T^{(d)}(z,\eta-|\zeta|)\right] + n_V\left[\hat{f}_V^{(d)}(z) + k_V^{(d)}(z,\eta-|\zeta|)\right]$$

$$+\tilde{n}_T\, g_T^{(d)}(z,\eta,|\zeta|) + \tilde{n}_V\, g_V^{(d)}(z,\eta,|\zeta|)$$

$$= e^{|\zeta|-\eta-z}\left[n_T\, f_T^{(d+1)}(z,\eta-|\zeta|) + n_V\, f_V^{(d+1)}(z,\eta-|\zeta|)\right]$$

$$+\,\tilde{n}_T\, g_T^{(d)}(z,\eta,|\zeta|) + \tilde{n}_V\, g_V^{(d)}(z,\eta,|\zeta|). \tag{4.50}$$

Note that p is an even function of ζ, as follows from T-duality $R_d \to 1/R_d$. n_T is the number of massless states for generic R_d, while n_V is the difference between the numbers of bosons and fermions which are massless. $\tilde{n}_T$ is the number of additional massless states at the self-dual point $R_d = 1$ and $\tilde{n}_V = \tilde{n}_T$ because these modes are bosons. Physically, the KK reduction of the 10-dimensional metric tensor along $S^1(R_d)$ provides generically an $U(1)$ gauge theory, which is enhanced to $SU(2)$ when $(R_d - 1/R_d)$, interpreted as a Higgs VEV, vanishes. The properties of the functions $g_T^{(d)}$ and $g_V^{(d)}$ is precisely to interpolate between the different massless spectra ($U(1)$ *versus* $SU(2)$). They are not functions of two complex structures z and $\eta - |\zeta|$ only, since the string scale $\sqrt{\alpha'}$ is entering the game. The definitions of the various functions appearing in Eq. (4.50) are

$$\hat{f}_T^{(d)}(z) = \frac{\Gamma\left(\frac{d+1}{2}\right)}{\pi^{\frac{d+1}{2}}}\sum_{\tilde{k}_0,\tilde{k}_9}\frac{e^{dz}}{\left[e^{2z}(2\tilde{k}_0+1)^2+(2\tilde{k}_9)^2\right]^{\frac{d+1}{2}}},$$

$$k_T^{(d)}(z,\eta-|\zeta|) = \sum_{m_d}{}' |m_d|^{\frac{d+1}{2}}\, e^{\frac{d+1}{2}(\eta-|\zeta|)}e^{dz}$$

$$\sum_{\tilde{k}_0,\tilde{k}_9}\frac{2K_{\frac{d+1}{2}}\left(2\pi|m_d|e^{\eta-|\zeta|}\sqrt{e^{2z}(2\tilde{k}_0+1)^2+(2\tilde{k}_9)^2}\right)}{\left[e^{2z}(2\tilde{k}_0+1)^2+(2\tilde{k}_9)^2\right]^{\frac{d+1}{4}}},$$

$$g_T^{(d)}(z,\eta,|\zeta|) = \left(e^{2|\zeta|}-1\right)^{\frac{d+1}{2}} e^{\frac{d+1}{2}(\eta-|\zeta|)}e^{dz}$$

$$\sum_{\tilde{k}_0,\tilde{k}_9}\frac{2K_{\frac{d+1}{2}}\left(2\pi(e^{2|\zeta|}-1)e^{\eta-|\zeta|}\sqrt{e^{2z}(2\tilde{k}_0+1)^2+(2\tilde{k}_9)^2}\right)}{\left[e^{2z}(2\tilde{k}_0+1)^2+(2\tilde{k}_9)^2\right]^{\frac{d+1}{4}}},$$

$$f_T^{(d+1)}(z,\eta-|\zeta|) = \frac{\Gamma\left(\frac{d}{2}+1\right)}{\pi^{\frac{d}{2}+1}}$$

$$\sum_{\tilde{k}_0,\tilde{k}_9,\tilde{m}_d}\frac{e^{(d+1)z}}{\left[e^{2z}(2\tilde{k}_0+1)^2+(2\tilde{k}_9)^2+e^{-2(\eta-|\zeta|)}\tilde{m}_d^2\right]^{\frac{d}{2}+1}}, \tag{4.51}$$

with the remaining ones given as

$$\begin{aligned} \hat{f}_V^{(d)}(z) &= e^{(d-1)z}\,\hat{f}_T^{(d)}(-z),\, k_V^{(d)}(z,\eta-|\zeta|) = e^{(d-1)z} \\ &\quad k_T^{(d)}(-z,\eta-|\zeta|+z), \end{aligned} \quad (4.52)$$

$$\begin{aligned} g_V^{(d)}(z,\eta,|\zeta|) &= e^{(d-1)z}\,g_T^{(d)}(-z,\eta+z,|\zeta|),\, f_V^{(d+1)}(z,\eta-|\zeta|) = e^{dz} \\ &\quad f_T^{(d+1)}(-z,\eta-|\zeta|+z). \end{aligned} \quad (4.53)$$

In the type II case, the partition function takes formally the form of the heterotic one, with $\tilde{n}_T = \tilde{n}_V = 0$. This is due to the fact that in type II, there is no enhancement of symmetry at $R_d = 1$.

Treating the circle $S^1(R_d)$ as an internal direction, the dimensional reduction from 10 to d dimensions involves the metric and dilaton field ϕ_d in d dimensions, together with the scalars η and ζ,

$$S = \int d^d x \sqrt{-g}\left(\frac{R}{2} - \frac{1}{2}(\partial\Phi)^2 - \frac{1}{2}(\partial\phi_\perp)^2 - \frac{1}{2}(\partial\zeta)^2 - \mathcal{F}\right), \quad (4.54)$$

where we have defined the normalized fields

$$\Phi := \frac{2}{\sqrt{(d-2)(d-1)}}\,\phi_d - \sqrt{\frac{d-2}{d-1}}\,\eta\,, \qquad \phi_\perp := \frac{2}{\sqrt{d-1}}\,\phi_d + \frac{1}{\sqrt{d-1}}\,\eta\,. \quad (4.55)$$

The 1-loop free energy density $\mathcal{F} = -Z_{1-\text{loop}}/(\beta V_{\text{box}})$ depends on the temperature T, the supersymmetry breaking scale M (*i.e.* Φ), ζ and implicitly on $\phi_\perp$ via η,

$$\mathcal{F} = -T^d\, p(z,\eta,\zeta)\,, \quad e^z = \frac{M}{T}\,, \quad T = \frac{e^{\frac{2\phi_d}{d-2}}}{2\pi R_0}\,, \quad M = \frac{e^{\frac{2\phi_d}{d-2}}}{2\pi R_9} \equiv \frac{e^{\sqrt{\frac{d-1}{d-2}}\Phi}}{2\pi}. \quad (4.56)$$

As usually, a FLRW ansatz for the metric and time-dependent scalars yields a stress-tensor whose thermal energy density and pressure are

$$P = T^d\, p(z,\eta,\zeta) \qquad \rho = T^d\, r(z,\eta,\zeta)\,, \quad r = (d-1)p - p_z. \quad (4.57)$$

Among the five independent equations of motions, we are particularly interested in the equation for ζ,

$$\ddot{\zeta} + (d-1)H\dot{\zeta} - \frac{\partial P}{\partial \zeta} = 0, \quad (4.58)$$

whose potential $-P$, as a function of ζ (the over fields held fixed), is shown on Figure 3 (when $z < 0$ *i.e.* $R_0 < R_9$). In the heterotic case, the profile of $-P$ can be divided in five phases, while in type II models the range I is reduced to a single point. The behavior in phase III (phase V) is exponentially decreasing (increasing), while when $z > 0$ and large enough, it is exponentially increasing (decreasing).

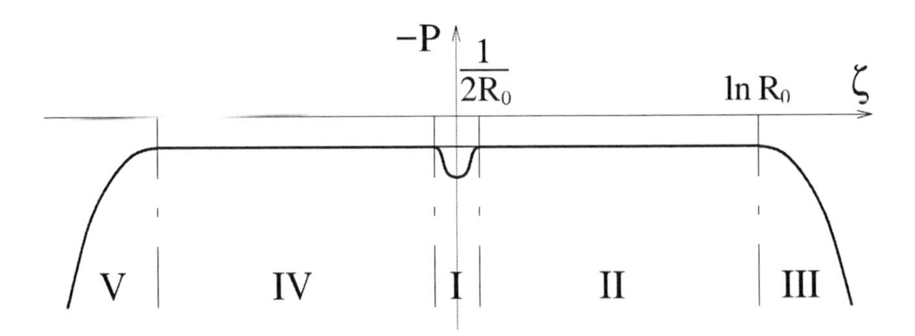

Figure 3: Qualitative shape of the thermal effective potential $-P$ of $\zeta = \ln R_d$ (the other variables held fixed, with $z < 0$). There are five phases in the heterotic models, while the range I is reduced to a single point in type II. When $z > 0$, one has to replace R_0 by R_9 in the boundaries of the ranges and if z is large enough, phase III (V) is increasing (decreasing).

Heterotic models

- **I : Higgs phase,** $\left|R_d - \dfrac{1}{R_d}\right| < \dfrac{1}{R_0}$ and/or $\dfrac{1}{R_9}$.

  The functions $k_T^{(d)}$ and $k_V^{(d)}$ in the first expression of Eq. (4.50) are exponentially suppressed and can be neglected. Since p is an even function of ζ, we have at the origin $p_\zeta = 0$ so that $\zeta(t) \equiv 0$ is a solution to (4.58). It follows that the pressure is drastically simplified, since

$$p(z, \eta, 0) = (n_T + \tilde{n}_T)\, \hat{f}_T^{(d)}(z) + (n_V + \tilde{n}_V)\, \hat{f}_V^{(d)}(z) := \tilde{p}(z)\,, \tag{4.59}$$

  i.e. does not depend on η. Thus, the analysis of the particular solution $\zeta \equiv 0$ brings us back to the study of section 3.1., once generalized in d dimensions. We conclude that when

$$-\frac{1}{2^d - 1} < \frac{n_V + \tilde{n}_V}{n_T + \tilde{n}_T} < 0, \tag{4.60}$$

  there exists an RDS^d (Radiation Dominated Solution in d dimensions). The behavior of the first order fluctuations around this solution are found to be exponentially damped. As a result, the RDS^d is a local attractor of the dynamics and R_d is stabilized at the self-dual point of enhanced symmetry.

- **II : Flat potential phase,** $\dfrac{1}{R_0}$ and $\dfrac{1}{R_9} < R_d - \dfrac{1}{R_d} < R_0$ and R_9.

  As in phase I, $k_T^{(d)}$ and $k_V^{(d)}$ in Eq. (4.50) can be neglected. Indeed, by definition, phase II starts when the functions $g_T^{(d)}$ and $g_V^{(d)}$ responsible for the interpolation between the generic and the enhanced massless spectra are exponentially suppressed as well. It follows that p is independent of η and ζ,

$$p(z, \eta, \zeta) \simeq n_T\, \hat{f}_T^{(d)}(z) + n_V\, \hat{f}_V^{(d)}(z) := \hat{p}(z)\,. \tag{4.61}$$

Consequently, any $\zeta(t) \equiv \zeta_0$ when ζ is in the range II solves Eq. (4.58). For any given ζ_0, we are back again to the analysis of section 3.1. and a particular RDS^d exists when

$$-\frac{1}{2^d - 1} < \frac{n_V}{n_T} < 0. \tag{4.62}$$

Small perturbations around such an RDS^d are found to the damped, even if the potential for ζ is flat. Actually, the fluctuations of ζ around any ζ_0 are suppressed due to the presence of "gravitational friction" in (4.58). In this sense, one can conclude that ζ is marginally stabilized.

- **III : Higher dimensional phase,** R_0 and/or $R_9 < R_d$.

As in phase II, $g_T^{(d)}$ and $g_V^{(d)}$ in Eq. (4.50) can be neglected. However, by definition, phase III starts when the functions $k_T^{(d)}$ and/or $k_V^{(d)}$ cease to be exponentially suppressed. In particular, when $R_d \gg R_0$ and R_9, one finds, using the second form of Eq. (4.50),

$$\begin{aligned} p(z,\eta,\zeta) \simeq e^{|\zeta|-\eta-z} \Big(& n_T\, \hat{f}_T^{(d+1)}(z) + n_V\, \hat{f}_V^{(d+1)}(z) \\ & + e^{d(z+\eta-|\zeta|)}(n_T + n_V)\frac{S_d^{oe}}{4} \Big) \,, \end{aligned} \tag{4.63}$$

where $S_d^{oe} = \frac{\Gamma(\frac{d}{2})}{\pi^{\frac{d}{2}}} \sum_m{}' \frac{1}{|m|^d}$. The term $e^{d(z+\eta-|\zeta|)} = (R_0/R_d)^d$ is power-like subdominant and exponentially small terms have been ignored.

Neglecting the small contribution $(R_0/R_d)^d$, the appearance of the functions $\hat{f}_T^{(d+1)}$ and $\hat{f}_V^{(d+1)}$ in p indicate that it is more natural to reconsider the system from a $(d+1)$-dimensional point of view. Regarding $S^1(R_4)$ as part of the space-time, R_4 is no longer an internal modulus but a component of the metric, $g_{st,dd} = (2\pi R_d)^2$. Denoting with primes all quantities in $d+1$ dimensions, the vacuum-to-vacuum energy density in the effective action in higher dimension, $Z_{1-\text{loop}}/(\beta V_{\text{box}} 2\pi R_d)$, is giving rise to can a pressure $P' = T'^{(d+1)} p'(z,\eta,\zeta)$, where

$$p'(z,\eta,\zeta) \simeq n_T\, \hat{f}_T^{(d+1)}(z) + n_V\, \hat{f}_V^{(d+1)}(z) := \hat{p}'(z)\,. \tag{4.64}$$

One more time, one concludes from the analysis of section 3.1. that an RDS^{d+1} with isotropic metric exists if

$$-\frac{1}{2^{d+1} - 1} < \frac{n_V}{n_T} < 0. \tag{4.65}$$

By isotropic metric, we mean that

$$e^{\xi} := b/a' = R_d/R_{\text{box}}, \tag{4.66}$$

where a' is the scale factor in the directions $1,\ldots,d-1$ and b is the scale factor in the direction d, satisfies $\xi(t) \equiv \xi_0$, a constant determined by the I.B.C.. Small perturbations around such a solution are shown to converge to zero. Since $\xi(t)$ measures the anisotropy of the local metric, one concludes that the attracting RDS^{d+1} is characterized by an enhanced spatial rotation group $SO(d-1) \to SO(d)$. The ratio e^{ξ_0} is interpreted as a "complex structure" of the "external space" we live in.

On the contrary, whenever the contribution $(R_0/R_d)^d$ in Eq. (4.63) is not negligible, we find it yields a "residual force" such that even though $\zeta(t)$ *i.e.* $R_d(t)$ is increasing, it is always caught by $R_9(t)$ and $R_0(t)$. Thus, the dynamics exits phase III and the system enters into region II, where the solution is attracted to an RDS ^{d}.

- **IV : Dual flat potential phase,** $\frac{1}{R_0}$ and $\frac{1}{R_9} < \frac{1}{R_d} - R_d < R_0$ and R_9.

  This region is the T-dual of phase II and has the same behavior, with $\zeta \to -\zeta$. The light states that contribute to $\mathbb{Z}_{1-\text{loop}}$ are the windings modes along $S^1(R_d)$ instead of the KK excitations.

- **V : Dual higher dimensional phase,** R_0 and/or $R_9 < \frac{1}{R_d}$.

  This region is the T-dual of phase III and has the same behavior, with $\zeta \to -\zeta$.

To summarize, when the Kähler modulus $R_d(t)$ is internal, it is attracted to the intersection of the ranges $1/R_i(t) < R_d(t) < R_i(t)$ $(i = 0, 9)$ and ends by being marginally stabilized or stabilized at $R_d = 1$. On the contrary, when $R_d(t)$ is large enough, the Universe is $(d+1)$-dimensional and R_d expands and runs away, with $R_d(t) \propto R_{\text{box}}(t)$. It is then better understood in terms of the complex structure R_d/R_{box}, which is marginally stabilized.

Type II superstring models

As said before, there is no enhanced symmetry point at $R_d = 1$ in the type II superstring models. Thus, their analysis can be derived from the heterotic one by taking $\tilde{n}_T = \tilde{n}_V = 0$. Since the local minimum of $-P$ at $\zeta = 0$ is not present anymore, the plateaux II and IV on Figure 3 are connected.

However, we expect by heterotic-type II duality that an Higgs phase I should exist in type II superstring at the non-perturbative level. A possible setup to describe this effect is to consider a pair of D-branes, whose distance is dual to the modulus R_d. In this context, our Universe is a "brane-world", whose spatial directions are parallel to the D-branes. The stabilization of R_d at the self-dual point on the heterotic side should imply the non-perturbative thermal effective potential in type II to force the D-branes to stay on top of each other, thus producing an $U(1) \to SU(2)$ enhancement. However, this attraction between the D-branes should only be local, since if they are separated enough so that the dual modulus R_d enters phase II, the thermal effective potential should allow stable finite distances between the D-branes. If the distance between the D-branes is very large, the force between them will be repulsive and the expanding Universe develops one more dimension.

An alternative non-perturbative type II set up realizing a gauge group enhancement involves singularities in the internal space. For example, a type IIA D2-brane wrapped on a vanishing 2-sphere whose radius is dual to R_d produces an $SU(2)$ gauge theory. It also admits a mirror description in type IIB [22]. The equivalence between the brane-world and geometrical singularity pictures can be analyzed along the lines of Ref. [23].

5. Conclusions and Perspectives

In this review, we describe some basics of early Universe cosmology in the framework of string theory. We first place the $(d-1)$-dimensional space in a very large box, while much smaller compactified directions span an internal space. To introduce temperature, we consider the Euclidean version of the background, where appropriate boundary conditions are imposed along the Euclidean time circle. Supersymmetry breaking is implemented in a similar way by choosing non-trivial boundary conditions along internal compact directions. In four dimensions, we analyze models where $\mathcal{N} = 4, 2, 1$ supersymmetry is spontaneously broken. The advantage of string theory is to cure all UV diverges encountered in field theory.

In this work, we restrict our study to the intermediate times $t_E \ll t \ll t_W$ *i.e.* after the end of the Hagedorn era and before the electroweak phase transition. The pressure and energy density of the gas of strings are computed from a microscopic point of view, using the 1-loop Euclidean string partition function. We then study the solutions of the low energy effective action and find that the quasi-static evolutions are attracted to Radiation Dominated Solutions. During the convergence to an RDS, there is no inflation (or a very tiny amount). However, interesting effects can occur. Internal radii that are participating into the spontanoues breaking of supersymmetry can be decompactified dynamically and lead to a change in the dimension of space-time. Moreover, the internal radii that are not participating in the spontaneous breaking of supersymmetry are stabilized. In general, the Universe tends to increase spontaneously its symmetries: The gauge symmetries or the local isotropy. Usually, the field theory description of the full evolution is in term of a succession of *different* field theories. The underlying string theory is required to connect them.

To go further, lots of work is still needed to unravel some experimental predictions from our framework. In four dimensions, models with spontaneously broken $\mathcal{N} = 1$ supersymmetry are particularly interesting, since they can include chiral matter. Dealing with them is under progress, in order to go beyond t_W. To do that, one needs to compute the radiative corrections to the different fields entering the action. Only after this full work is accomplished, it becomes possible to discuss what happens in the matter dominated era, and observe if a late time inflation era can exist. There are strong beliefs that $\mathcal{N} = 1$ models will produce a non zero cosmological constant.

Another direction of work concerns the relaxation of the homogeneity assumption. We could thus deal with the issues of entropy production and adiabaticity. On simpler grounds, it could be interesting to show that small initial homogeneities disappear and that the standard FLRW ansatz is an attractor in the space of backgrounds.

Finally, an important issue to examine is the existence of *a mechanism alternative to inflation* during the Hagedorn era. There are already some proposals in this direction in Ref. [16]. Preliminary results by some of the authors indicate that such a mechanism is plausible for non-pathological sting vacua, where the Hagedorn transition is resolved [18, 19].

Acknowledgments

The authors would like to thank NOVA publishers for giving us the opportunity to write this review. We are grateful to N. Toumbas for useful discussions. H.P. thanks the Ecole

Normale Supérieure for hospitality.
This work is partially supported by the ANR (CNRS-USAR) contract 05-BLAN-0079. The work of F.B, J.E. and H.P is supported by the European contracts PITN-GA-2009-237920, ERC-2008-AdG 20080228, the CNRS PICS contracts 3747, 4172 and the INTAS grant 03-51-6346. J.E. acknowledges financial support from the Groupement d'Int érêt Scientifique P2I.

References

[1] C. Kounnas and H. Partouche, "Instanton transition in thermal and moduli deformed de Sitter cosmology," Nucl. Phys. B **793** (2008) 131 [arXiv:0705.3206 [hep-th]];

C. Kounnas and H. Partouche, "Inflationary de Sitter solutions from superstrings," Nucl. Phys. B **795** (2008) 334 [arXiv:0706.0728 [hep-th]].

[2] T. Catelin-Jullien, C. Kounnas, H. Partouche and N. Toumbas, "Thermal/quantum effects and induced superstring cosmologies," Nucl. Phys. B **797** (2008) 137 arXiv:0710.3895 [hep-th];

T. Catelin-Jullien, C. Kounnas, H. Partouche and N. Toumbas, "Thermal and quantum superstring cosmologies," Fortsch. Phys. **56** (2008) 792 [arXiv:0803.2674 [hep-th]].

[3] T. Catelin-Jullien, C. Kounnas, H. Partouche and N. Toumbas, "Induced superstring cosmologies and moduli stabilization," arXiv:0901.0259 [hep-th].

[4] F. Bourliot, C. Kounnas and H. Partouche, "Attraction to a radiation-like era in early superstring cosmologies," Nucl. Phys. B **816** (2009) 227 [arXiv:0902.1892 [hep-th]];

F. Bourliot, J. Estes, C. Kounnas and H. Partouche, "Cosmological Phases of the String Thermal Effective Potential," arXiv:0908.1881 [hep-th].

[5] J. Estes, C. Kounnas and H. Partouche, "Superstring cosmology for $\mathcal{N}_4 = 1 \to 0$ superstring vacua," LPTENS–09/32, CPHT–RR080.0709 in print.

[6] T. Morii, C.S. Lim, S.N. Mukherjee, "The physics of the standard model and beyond," (World Scientific, 2004).

[7] J. Polchinski, "String Theory," **Vol. 1** & **2**, (Cambridge University press, 2001).

[8] S. Perlmutter, "A study of 42 type Ia supernovae and a resulting measurement of Omega(M) and Omega(Lambda)," Physics Reports-Review section of Physics Letters **307 (1-4)**: 325-331 (1998).

[9] P.J.E. Peebles, "Principles of Physical Cosmology," (Princeton series in Physics, 1993).

[10] R.M. Wald, "General Relativity," (The University of Chicago Press, 1984).

[11] F. Bourliot, P. G. Ferreira, D. F. Mota and C. Skordis, "The cosmological behavior of Bekenstein's modified theory of gravity," Phys. Rev. D **75** (2007) 063508 [arXiv:astro-ph/0611255].

[12] H. Zhao, "Constraining TeVeS gravity as effective dark matter and dark energy," Int. J. Mod. Phys. D **16** (2008) 2055 [arXiv:astro-ph/0610056].

[13] R. Hagedorn, "Statistical thermodynamics of strong interactions at high-energies," Nuovo Cim. Suppl. **3**, 147 (1965);

K. Huang and S. Weinberg, "Ultimate temperature and the early universe," Phys. Rev. Lett. **25**, 895 (1970);

B. Sathiapalan, "Vortices on the string world sheet and constraints on toral compactification," Phys. Rev. D **35**, 3277 (1987);

Y. I. Kogan, "Vortices on the world sheet and string's critical dynamics," JETP Lett. **45**, 709 (1987) [Pisma Zh. Eksp. Teor. Fiz. **45**, 556 (1987)];

M. Axenides, S. D. Ellis and C. Kounnas, "Universal behavior of D-dimensional superstring models," Phys. Rev. D **37**, 2964 (1988);

D. Kutasov and N. Seiberg, "Number of degrees of freedom, density of states and tachyons in string theory and CFT," Nucl. Phys. B **358** (1991) 600.

[14] J. Atick and E. Witten, "The Hagedorn transition and the number of degrees of freedom of string theory," Nucl. Phys. B **310**, 291 (1988).

[15] C. Kounnas and B. Rostand, "Coordinate dependent compactifications and discrete symmetries," Nucl. Phys. B **341** (1990) 641.

[16] A. Nayeri, R. H. Brandenberger and C. Vafa, "Producing a scale-invariant spectrum of perturbations in a Hagedorn phase of string cosmology,' Phys. Rev. Lett. **97**, 021302 (2006) [arXiv:hep-th/0511140];

R. H. Brandenberger, A. Nayeri, S. P. Patil and C. Vafa, "Tensor modes from a primordial Hagedorn phase of string cosmology," Phys. Rev. Lett. **98**, 231302 (2007) [arXiv:hep-th/0604126];

R. H. Brandenberger, A. Nayeri, S. P. Patil and C. Vafa, "String gas cosmology and structure formation," Int. J. Mod. Phys. A **22**, 3621 (2007) [arXiv:hep-th/0608121];

R. H. Brandenberger, "Alternatives to cosmological inflation," arXiv:0902.4731 [hep-th].

[17] I. Antoniadis and C. Kounnas, "Superstring phase transition at high temperature," Phys. Lett. B **261** (1991) 369;

I. Antoniadis, J. P. Derendinger and C. Kounnas, "Non-perturbative supersymmetry breaking and finite temperature instabilities in $\mathcal{N} = 4$ superstrings," arXiv:hep-th/9908137;

C. Kounnas, "Universal thermal instabilities and the high-temperature phase of the $\mathcal{N} = 4$ superstrings," arXiv:hep-th/9902072.

[18] C. Angelantonj, C. Kounnas, H. Partouche and N. Toumbas, "Resolution of Hagedorn singularity in superstrings with gravito-magnetic fluxes," Nucl. Phys. B **809** (2009) 291 [arXiv:0808.1357 [hep-th]].

[19] C. Kounnas, "Massive boson–fermion degeneracy and the early structure of the universe," Fortsch. Phys. **56** (2008) 1143 [arXiv:0808.1340 [hep-th]];

I. Florakis and C. Kounnas, "Orbifold symmetry reductions of massive boson-fermion degeneracy," Nucl. Phys. B **820** (2009) 237 [arXiv:0901.3055 [hep-th]].

[20] J. R. Ellis, C. Kounnas and D. V. Nanopoulos, "No scale supersymmetric GUTs," Nucl. Phys. B **247** (1984) 373;

J. R. Ellis, C. Kounnas and D. V. Nanopoulos, "Phenomenological $SU(1,1)$ supergravity," Nucl. Phys. B **241** (1984) 406;

J. R. Ellis, A. B. Lahanas, D. V. Nanopoulos and K. Tamvakis, "No-scale supersymmetric standard model," Phys. Lett. B **134**, 429 (1984).

[21] J. Scherk and J. H. Schwarz, "Spontaneous breaking of supersymmetry through dimensional reduction," Phys. Lett. B **82** (1979) 60;

R. Rohm, "Spontaneous supersymmetry breaking in supersymmetric string theories," Nucl. Phys. B **237** (1984) 553;

C. Kounnas and M. Porrati, "Spontaneous supersymmetry breaking in string theory," Nucl. Phys. B **310** (1988) 355;

S. Ferrara, C. Kounnas, M. Porrati and F. Zwirner, "Superstrings with spontaneously broken supersymmetry and their effective theories," Nucl. Phys. B **318** (1989) 75.

[22] A. Strominger, "Massless black holes and conifolds in string theory," Nucl. Phys. B **451** (1995) 96 [arXiv:hep-th/9504090];

S. Kachru and C. Vafa, "Exact results for $\mathcal{N} = 2$ compactifications of heterotic strings," Nucl. Phys. B **450** (1995) 69 [arXiv:hep-th/9505105];

I. Antoniadis and H. Partouche, "Exact monodromy group of $\mathcal{N} = 2$ heterotic superstring," Nucl. Phys. B **460** (1996) 470 [arXiv:hep-th/9509009];

H. Partouche, "Non perturbative check of $\mathcal{N} = 2$, $D = 4$ heterotic/type II duality," Nucl. Phys. Proc. Suppl. **55B** (1997) 210 [arXiv:hep-th/9610119].

[23] P. Kaste and H. Partouche, "On the equivalence of $\mathcal{N} = 1$ brane worlds and geometric singularities with flux," JHEP **0411** (2004) 033 [arXiv:hep-th/0409303].

In: Superstring Theory in the 21st Century
Editor: Gerold B. Charney, pp. 161-186
ISBN: 978-1-61668-385-6

Chapter 6

SUPERSYMMETRIC STANDARD MODEL AND ITS PARTICLE SPECTRUM FROM FOUR DIMENSIONAL SUPERSTRING*

B. B. Deo and L. Maharana**
Department of Physics, Utkal University, Bhubaneswar-751004, India

Abstract

Nambu-Goto [2] proposed a simple open string as a sequel to the dual resonance model. Afterwards, Lovelace [4] had shown that the physical states have negative norms except in 26 dimension. The discovery of Scherck and Schwartz [5] that the spectrum contains graviton, raised the ultimate hope that all the four interactions, observed in nature can be unified. But there are also the unphysical tachyons which were subsequently found to vanish in ten dimensional superstring constructed by Green and Schwartz [6]. Then came the phenomenological rise in both conceptual and mathematical complexity and the simplicity of the Nambu-Goto string was forgotten. This paper and few others [14]-[16] attempt at a successful revival of the classic idea. First of all, noting the Mandelstam's equivalence of one bosonic mode to two fermionic modes, the Nambu-Goto string in 26 dimension is written as the four coordinates and 4 groups of eleven fermions in SO(3,1) bosonic representations. It is proved that the world sheet action is supersymmetric, rich in particle quanta and is equivalent to a four dimensional superstring. Constructing the supersymmetric charge, the Hamiltonian is found to allow only positive energy states, proving that there are no tachyons. There is also manifest modular invariance.

At first, there appears to be many ghosts and negative norm states. After quantising in NS and R formulations, we construct the Virasoro operators, the superalgebra and the physical states. The BRST supercharge, found by using conformal and superconformal ghosts, is nilpotent. The Theory is ghost free, anomaly free and is unitary.

With a definite proof that the Nambu-Goto string in a superstring formulation is correct, we see that the symmetry of action $SO(6) \otimes SO(5)$ is also the gauge symmetry and can descend to three generations of Supersymmetric standard model by the

*A version of this chapter was also published in New Developments in String Theory Research, edited by Susan A. Grece published by Nova Science Publishers. It was submitted for appropriate modifications in an effort to encourage wider dissemination of research. **E-mail address: lmaharan@iopb.res.in

use of Wilson loop. For completeness, we construct the particle spectrum, consisting of the Higgs sector, Gauge sector and the fermions in detail. As an interesting addition, we show that the particles of the standard model are superpartners of each other, following Nambu and Witten. From a simple Nambu-Goto superstring action, we construct the graviton and gravitino states which was first ever written down action of N=1 supergravity.

1 Introduction

From 1960 onward, the high energy accelerators poured down enormous amount of data on the properties of hadrons. Trying to systematize them, there was success in putting all the masses in the Regge trajectories. These trajectories were rising linearly and could have possibly risen to infinite and this have been verified till the angular momentum J= $\frac{11}{2}$. In the meantime, Veneziano [1] suggested an elegant equation which was fully crossing symmetric. The dual model of strong interaction took shape. This provided the motivation for Nambu and Goto[2] to write down the equation of motion of a string in analogy to a relativistic point particle and proposed that the excitations of this string result in hadronic quanta. This was written in first quantised form and later it was transformed in to the second quantised form following Gervais and Sakita [3]. String theory took quantum shape. Soon afterwards, Lovelace [4] found that the norm of the physically admissible states depend on the dimensions on which the string lives. The bosonic string makes sense only in 26 dimensions.

Dual theory of strong interaction gained momentum. Unfortunately, all the particles could not be incorporated and a large number of massless particles showed up with arbitrary spin. Astonishingly, the closed string of the dual model turned out to have massless spin two quanta. In 1974, Scherck and Schwartz [5] made the landmark suggestion, that this quanta is, indeed, a graviton. There appeared a distinct possibility of unifying the four interactions namely, the strong, the electromagnetic, the weak and the gravitational, by developing the theory of string. Such a theory was named as the theory of everything (TOE). There was a real hope that a consistent theory of gravity would energy sooner than expected from string theory.

There was a difficulty in a way that the tachyons were found. To remedy this and taking the advantage of the development of the supersymmetric theories, the superstrings were constructed but they also live in ten dimensions. Even though, Green and Schwartz [6] have shown that the ten dimensional superstring is anomaly free and $E_8 \otimes E_8$ "heterotic string" of Gross, Harvey and Martinec [7] considered the best model to proceed to unify gravity with physically reasonable compactified models of particle interactions, the simplicity of Nambu-Goto string in four dimensions got lost. This article is an attempt to construct an equivalent superstring theory to revive the simplicity and wider applicability of four dimensional superstring.

A lot of research work has been done to construct a four dimensional string [8], specifically in the later half of the eighties. Antoniadis et al [9] have constructed the theory using eighteen real fermions in trilinear couplings. Kawai et al [10] have constructed varieties of models in four dimensions. None of them have noted the relevance of the strings to the Standard Model physics. We record the works of Gates and his collaborators [11], whose contents are similar to ours to some extent.

Consistent superstrings as solution of D=26 bosonic string has been shown to exist by Casper,Englert, Nicolai and Taormina [12]. Fairly recent is the extension of this work by Englert, Hourt and Taormina [13] to brane fusion in the bosonic string resulting in the fermionic string. The present work which has similar purpose, is quite different. We search for a way to construct a superstring within a Nambu-Goto string in 26 dimensions by introducing a eleven four-plot Majorana fermions in the vector representation(bosonic) of Lorentz group SO(3,1) and arrive, finally at a four dimensional but not a ten dimensional superstring.

This article is organised in the following sequence. The details of novel string is given in Section 2 as contained in Reference [14]. In Section 3, the local two dimensional supersymmetric extension is given and the current and energy-momentum constraints are deduced by the usual variational method. Since the superstring can not have tachyons, we construct the world sheet supersymmetric charge and show in this Section 4 that the Hamiltonian cannot have negative values.

In Section 5, we quantise one of the two equivalent actions, which has a rich particle spectrum. The super Virasoro generators are constructed and are shown to satisfy super Virasoro algebra with the usual anomaly terms. The physical states are defined and the bosonic and the fermionic null states are constructed in Section 6 to foresee the elimination of ghosts from the Fock space. In Section 7, a detail discussion is given on the conformal and the superconformal ghosts. The nilpotent BRST charge is constructed, thereby proving that the theory is unitary and is ghost and anomaly free. A detailed discussion as to how the theory is modular invariant is given in Section 8.

The theory as such would not have had any fruitful relevance except its immediate applicability to low energy particles, as embedded in Supersymmetric Standard Model(SSSM). In Section 9, we show that the symmetry of action $SO(6) \otimes SO(5)$ is also the gauge symmetry. Without breaking supersymmetry, it can descend to three generations of the Standard Model objects. In Section 10, we also construct the zero mass vector bosons and Higgs in the bosonic sector and massless quarks and leptons in the fermionic sector [15]. In Section 11, we give an illuminating proof that bosons and fermions of MSSM are superpartners of each other, related by supersymmetric transformations.

Finally, we conclude the article by finding the simplest supergravity multiplet from the Nambu-Goto 4-dimensional superstring. We would like to make a special mention of the construction of Einstein's equation of general relativity as worked out by us in reference [16].

2 Construction of Superstring in Four Dimensions

Nambu-Goto string in 26 dimension has a simple form. It is given by the action in the world sheet (σ, τ)

$$S = -\frac{1}{2\pi} \int d^2\sigma \, \partial^\alpha X^\mu(\sigma, \tau) \, \partial_\alpha X_\mu(\sigma, \tau) \tag{1}$$

The α=0 or 1 refers to σ or τ. X^μ's are the twenty six bosonic Lorentz vector coordinates. The fermions can be introduced in the action (1) and reduce the bosonic coordinates to four only, by adding to the latter 44 fermions 'ϕ' as

$$S = -\frac{1}{2\pi} \int d^2\sigma \left[\partial^\alpha X^\mu(\sigma,\tau)\, \partial_\alpha X_\mu(\sigma,\tau) - i \sum_{j=1}^{44} \bar{\phi}^j \rho^\alpha \partial_\alpha \phi_j \right] \tag{2}$$

The two dimensional Dirac matrices are

$$\rho^0 = \begin{pmatrix} 0 & -i \\ i & 0 \end{pmatrix}, \quad \rho^1 = \begin{pmatrix} 0 & i \\ i & 0 \end{pmatrix} \tag{3}$$

and $\bar{\phi} = \phi^\dagger \rho^o$. The fermions are world sheet scalars with internal symmetry SO(44). To proceed to find a superstring action, we should find vectorial spinors. Fortunately, there are real Majorana fermions which are vectors in the bosonic representation of SO(3,1), without violating spin-statistics theorem. Calling these fermion vectors ψ^μ, the action (2) is improved to an SO(11) symmetric action

$$S = -\frac{1}{2\pi} \int d^2\sigma \left[\partial^\alpha X^\mu(\sigma,\tau)\, \partial_\alpha X_\mu(\sigma,\tau) - i \sum_{j=1}^{11} \bar{\psi}^{\mu,j} \rho^\alpha \partial_\alpha \psi_{\mu,j} \right] \tag{4}$$

We searched for linear combinations of ψ's to find a Ψ^μ to be the super partner of X^μ. After a lot of guessing and probing, we found that there is one viable possibility. The eleven vectors are further divided into species, $\psi^{\mu,j}$: j=1,2,..,6 and $\phi^{\mu,k}$: k=7,8,...,11. To have a correct anticommutation property of ψ^μ, for one group, the positive and negative frequency parts are $\psi^{\mu,j} = \psi^{+\mu,j} + \psi^{-\mu,j}$ where as for the other $\phi^{\mu,k} = \phi^{+\mu,k} - \phi^{-\mu,k}$ The latter is allowed for the massless fermions due to the uncertainty in the phase of creation operators. Finally we arrive at a $SO(6) \otimes SO(5)$ invariant action

$$S = -\frac{1}{2\pi} \int d^2\sigma \left[\partial^\alpha X^\mu(\sigma,\tau)\, \partial_\alpha X_\mu(\sigma,\tau) - i\bar{\psi}^{\mu,j} \rho^\alpha \partial_\alpha \psi_{\mu,j} + i\bar{\phi}^{\mu,k} \rho^\alpha \partial_\alpha \phi_{\mu,k} \right] \tag{5}$$

The upper index 'j' refers to a row and lower index 'j' to a column. Even though we shall be working with this action, we have still additional component indices j and k. To isolate and peel them off, we consider rows e^j and e^k, eleven numbered row matrices with the only nonvanishing element 'one' in the jth or k^{th} place. $e^j e_j$=6 and $e^k e_k$=5. To be specific, they look like

$$\begin{aligned} e^1 &= (1,0,0,0,0,0,0,0,0,0,0) && (6) \\ \textit{and}\ \ e^7 &= (0,0,0,0,0,0,-1,0,0,0,0) && (7) \end{aligned}$$

and it is to be noted that in the 11×11 matrix, the element $e^1 e^3$ does not occupy the same place as $e^3 e^1$ element.

Now we find that the action (5) is supersymmetric under the transformations

$$\begin{aligned} \delta X^\mu &= \bar{\epsilon}\left(e^j \psi^\mu_j - e^k \phi^\mu_k \right), && (8) \\ \delta \psi^{\mu,j} &= -i\epsilon\, e^j \rho^\alpha \partial_\alpha X^\mu, && (9) \\ \textit{and}\quad \delta \phi^{\mu,k} &= -i\epsilon\, e^k \rho^\alpha \partial_\alpha X^\mu. && (10) \end{aligned}$$

ϵ is a constant anticommuting spinor.

In action (5), there are 44 fermionic modes, but only 4 bosonic modes. If there is such mismatch, two successive supersymmetric transformations will not lead to a translation. One has to introduce additional auxilliary fields. In our case, in particular, the transformations lead to a spatial translation with the coefficient $a^\alpha = 2i\bar{\epsilon}^1\rho^\alpha\epsilon_2$, namely,

$$\begin{aligned} [\delta_1, \delta_2]\, X^\mu &= a^\alpha \partial_\alpha X^\mu, & (11)\\ [\delta_1, \delta_2]\, \psi^{\mu,j} &= a^\alpha \partial_\alpha \psi^{\mu,j}, & (12)\\ \textit{and} \quad [\delta_1, \delta_2]\, \phi^{\mu,k} &= a^\alpha \partial_\alpha \phi^{\mu,k} & (13) \end{aligned}$$

The results are true, only and only if

$$\psi_j^\mu = e_j \Psi^\mu, \qquad \phi_k^\mu = e_k \Psi^\mu \tag{14}$$

with the linear combination,

$$\Psi^\mu = e^j \psi_j^\mu - e^k \phi_k^\mu. \tag{15}$$

It is easy to verify that

$$\begin{aligned} \delta X^\mu &= \bar{\epsilon}\Psi^\mu, & \delta\Psi^\mu &= -ie\rho^\alpha\partial_\alpha X^\mu & (16)\\ [\delta_1, \delta_2]\, X^\mu &= a^\alpha\partial_\alpha X^\mu, & [\delta_1, \delta_2]\, \Psi^\mu &= a^\alpha\partial_\alpha\Psi^\mu. & (17) \end{aligned}$$

Thus Ψ^μ in equation (15) is the superpartner of X^μ. Equation (14) effect the separation of the j, k indices with only one Ψ^μ in the jth or k^{th} place while in these specified sites they emit and absorb quanta as obtained by quantising ψ_j^μ and ϕ_k^μ of the action of equation (5). It is surprising and comes as bonus is that the action (5) reduces to a Nambu-Goto superstring action in four dimensions.

$$S = -\frac{1}{2\pi}\int d^2\sigma \left[\partial^\alpha X^\mu(\sigma,\tau)\, \partial_\alpha X_\mu(\sigma,\tau) - i\bar{\Psi}^\mu\rho^\alpha\partial_\alpha\Psi_\mu\right] \tag{18}$$

Actions (5) and (18) are equivalent. The action (18) has equal number of bosons and fermions. We shall however concentrate on the $SO(6)\otimes SO(5)$ invariant action (5) which is rich in particle content and much more illuminating. As follows from action(5), the nonvanishing equal time commutator and anticommutators are [17]

$$\begin{aligned} \left[\partial_\pm X^\mu(\sigma,\tau), \partial_\pm X^\nu(\sigma',\tau)\right] &= \pm\frac{\pi}{2}\eta^{\mu\nu}\delta'(\sigma-\sigma') & (19)\\ \{\psi_A^\mu(\sigma,\tau), \psi_B^\nu(\sigma',\tau)\} &= \pi\eta^{\mu\nu}\delta(\sigma-\sigma')\delta_{AB} & (20)\\ \{\phi_A^\mu(\sigma,\tau), \phi_B^\nu(\sigma',\tau)\} &= -\pi\eta^{\mu\nu}\delta(\sigma-\sigma')\delta_{AB} & (21)\\ \{\Psi^\mu(\sigma,\tau), \Psi^\nu(\sigma',\tau)\} &= \pi\eta^{\mu\nu}\delta(\sigma-\sigma') & (22) \end{aligned}$$

We now use the following five important well known equations deduced in reference [17]. The Noether supercurrent is

$$J_\alpha = \frac{1}{2}\rho^\beta\rho_\alpha\Psi^\mu\partial_\beta X^\mu \tag{23}$$

The light cone components of the current and energy momentum tensors are, on the basis (ψ^+, ψ^-) and (ϕ^+, ϕ^-),

$$J_+ = \partial_+ X_\mu \Psi_+^\mu \quad (24)$$

$$J_- = \partial_- X_\mu \Psi_-^\mu \quad (25)$$

$$T_{++} = \partial_+ X^\mu \partial_+ X_\mu + \frac{i}{2}\bar{\psi}_+^{\mu,j}\partial_+\psi_{+\mu,j} - \frac{i}{2}\bar{\phi}_+^{\mu,k}\partial_+\phi_{+\mu,k} \quad (26)$$

$$T_{--} = \partial_- X^\mu \partial_- X_\mu + \frac{i}{2}\bar{\psi}_-^{\mu,j}\partial_-\psi_{-\mu,j} - \frac{i}{2}\bar{\phi}_-^{\mu,k}\partial_-\phi_{-\mu,k} \quad (27)$$

As usual, $\partial_\pm = \frac{1}{2}(\partial_\tau \pm \partial_\sigma)$. At equal time τ,

$$[T_{++}(\sigma), T_{++}(\sigma')] = i\pi\delta'(\sigma - \sigma')[T_{++}(\sigma) + T_{++}(\sigma')] \quad (28)$$

$$[T_{++}(\sigma), J_+(\sigma')] = i\pi\delta'(\sigma - \sigma')[J_+(\sigma) + \frac{1}{2}J_+(\sigma')] \quad (29)$$

To satisfy Jacobi Identity

$$\Big[T_{++}(\sigma), \{J_+(\sigma'), J_+(\sigma'')\}\Big] = \Big\{T_{++}(\sigma), [J_+(\sigma'), J_+(\sigma'')]\Big\} + (\sigma' \leftrightarrow \sigma'') \quad (30)$$

and using $\delta(\sigma - \sigma')\delta'(\sigma - \sigma'') + \delta'(\sigma - \sigma')\delta(\sigma - \sigma'') = \delta(\sigma' - \sigma'')\delta'(\sigma - \sigma')$, we verify that

$$\{J_+(\sigma), J_+(\sigma')\} = \pi\delta(\sigma - \sigma')T_{++}(\sigma) \quad (31)$$

$$\{J_-(\sigma), J_-(\sigma')\} = \pi\delta(\sigma - \sigma')T_{--}(\sigma) \quad (32)$$

$$\{J_+(\sigma), J_+(\sigma')\} = 0 \quad (33)$$

Thus the algebra closes and anticipates, in advance, the closure of super Virasoro algebra.

3 Local 2D Supersymmetry and the Constraints

First, we discuss, the nature and number of fermionic metric ghosts. Equations (20) and (21) suggest there are eleven of them. But equation (22), following from parent action, there are only two of them. The elevens follow from these two by multiplication with e^j or e^k as we shall see. Usually the constraints are obtained as postulates of vanishing of currents and the energy momentum tensor. This can be derived rigorously from a local 2d supersymmetric action. (X^μ, Ψ^μ) form a supersymmetric pair. In addition to this, we introduce another pair, a 'Zweibein' $e_\alpha(\sigma, \tau)$ and its supersymmetric partner, the gravitino $\chi_\alpha = \nabla_\alpha e$. The real 2d action is

$$S = -\frac{1}{2\pi}\int d^2\sigma e\left[h^{\alpha\beta}\partial_\alpha X^\mu \partial_\beta X_\mu - i\Psi^\mu \rho^\alpha \partial_\alpha \Psi_\mu + 2\bar{\chi}_\alpha \rho^\beta \rho^\alpha \Psi^\mu \partial_\beta X_\mu + \frac{1}{2}\Psi^\mu \Psi_\mu \bar{\chi}_\beta \rho^\beta \rho^\alpha \chi_\alpha\right] \quad (34)$$

This was first written down by Brink, Di Vecchia, Deser and Zumino [18]. This action (34) is invariant under local supersymmetric transformation,

$$\delta X^\mu = \bar{\epsilon}\bar{\Psi}^\mu; \quad \delta\Psi^\mu = -i\rho^\alpha\epsilon(\partial_\alpha X^\mu - \bar{\Psi}^\mu\chi_\alpha) \tag{35}$$

$$and \quad \delta\epsilon_\alpha = -2i\bar{\epsilon}\rho^\alpha\chi_\alpha, \quad \partial\chi_\alpha = \nabla_\alpha\epsilon \tag{36}$$

There are many other interesting transformation listed in detail in standard string theory text books.

For deriving the constraints, variation of the action with respect to e_α gives the vanishing of the energy momentum tensor while the variation with respect to χ_α gives J_α=0 with

$$J_\alpha = \frac{\pi}{2\epsilon}\frac{\delta S}{\delta\bar{\chi}^\alpha} = \rho^\beta\rho_\alpha\Psi^\mu\partial_\beta X_\mu - \frac{1}{4}\bar{\Psi}^\mu\Psi_\mu\rho_\alpha\rho^\beta\chi_\beta \tag{37}$$

But the gravitino field χ_α decouples and in the gauge χ_α=0, we have

$$J_\alpha = \rho^\beta\rho_\alpha\Psi^\mu\partial_\beta X_\mu = 0 \tag{38}$$

$$T_{\alpha\beta} = -\frac{\pi}{2\epsilon}e_{\alpha\gamma}T^\gamma_\beta = \partial_\alpha X^\mu\partial_\beta X^\mu - \frac{1}{2}\Psi^\mu\rho_{(\alpha}\partial_{\beta)}\Psi_\mu - (trace) = 0. \tag{39}$$

Thus, J_+, J_- and T_{++}, T_{--} as given earlier in equations (23) to (27), vanish. These are the four constraints and are sufficient. The eleven constraints follow, by noting that $\psi^{\mu,j} = e^j\Psi^\mu$ and $\phi^{\mu,k} = e^k\Psi^\mu$ and one can get all the component constraints from equations (24) and (25).

$$\partial_\pm X_\mu\psi^\mu_{\pm j} = \partial_\pm X^\mu e_j\Psi_{\pm\mu} = 0, \tag{40}$$

$$\partial_\pm X_\mu\phi^\mu_{\pm k} = \partial_\pm X^\mu e_k\Psi_{\pm\mu} = 0 \tag{41}$$

These are not really necessary except for remembering these fields are from the action (5).

4 World Sheet Supersymmetry Algebra and Absence of Tachyons

It is necessary criteria for a superstring that there should not be any unstable and physically unacceptable tachyons. One needs to examine carefully the restrictions on the Fock space due to supersymmetric algebra. One should be able to construct a world sheet supersymmetric charge 'Q' so that we reproduce the equation

$$\delta X^\mu = \left[X^\mu, \bar{Q}\epsilon\right] = \bar{\epsilon}\Psi^\mu \tag{42}$$

$$\delta\Psi^\mu = \left[\Psi^\mu, \bar{Q}\epsilon\right] = -i\rho^\alpha\partial_\alpha X^\mu\epsilon \tag{43}$$

Inspection shows that we should have

$$\bar{Q} = -\frac{i}{\pi}\int_0^\pi d\sigma\Psi_\mu\rho^\alpha\rho^0\partial_\alpha X^\mu \tag{44}$$

leading to

$$Q^\dagger = -\frac{i}{\pi}\int_0^\pi d\sigma \Psi_\mu \rho^\alpha \rho^0 \partial_\alpha X^\mu \tag{45}$$

and

$$Q = -\frac{i}{\pi}\int_0^\pi d\sigma \Psi_\mu \rho_0 \rho^{\alpha\dagger} \partial_\alpha X^\mu \tag{46}$$

It is a straight forward but lengthy calculation to find

$$\sum_\alpha \{Q_\alpha^\dagger, Q_\alpha\} = 2H \tag{47}$$

where H is the Hamiltonian of this superstring. It follows that for any state $|\phi_0>$ in the Fock space

$$\sum_\alpha |Q_\alpha|\Phi_0>|^2 = 2 < \phi_0|H|\phi_0> \;\; \geq 0 \tag{48}$$

As the Hamiltonian is the sum of squares, there cannot be any surviving tachyons in the physical space. After quantising and obtaining the mass spectrum we will find two tachyons, one in the fermionic and the other in bosonic sector. We shall show that the energies cancel as it must be due to supersymmetry and will not violate equation (48).

5 Quantisation and Virasoro Algebra Generators

The Brink et al [18] invariant action (34) under 2d-local transformations is not invariant under four dimensional local supersymmetric transformation. In this theory, it is simple to obtain the Green-Schwartz action for N=1 local supersymmetry. There are four component Dirac spinors in the representation of SO(3,1) which we denote by $\theta_{j,\delta}$ and $\theta_{k,\delta}$. δ is the real space time Dirac spinor component index running from 0 to 3. It is worthwhile to note that both SO(6) and SO(5) have four component fermion representations. A real space time fermion rather than a vector can be constructed as

$$\Theta_\alpha = \sum_{j=1}^{6} e^j \theta_{j\alpha} - \sum_{k=7}^{11} e^k \theta_{k\alpha} \tag{49}$$

The Green-Schwartz action is, in the four dimensions,

$$S = \frac{1}{2\pi}\int d^2\sigma \left(\sqrt{g} g^{\alpha\beta}\Pi_\alpha \Pi_\beta + 2i\epsilon^{\alpha\beta}\partial_\alpha X^\mu \bar{\Theta}\Gamma_\mu \partial_\beta \Theta\right) \tag{50}$$

where Γ_μ's are the 4x4 Dirac gamma matrices and

$$\Pi_\alpha^\mu = \partial_\alpha X^\mu - i\bar{\Theta}\Gamma^\mu \partial_\alpha \Theta. \tag{51}$$

This action is space time supersymmetric in four dimensions. Details can be found in reference [17].

The major defect with this action is that the standard covariant quantisation does not work. To proceed with the covariant formulation for quantising the particle-spectrum rich action (5) instead of the equivalent action (18), we have to implement NS-R [19] scheme with G.S.O. [20] projection which is simple, elegant and equivalent.

The bosonic coordinates in the open string are quantised as

$$X^\mu(\sigma,\tau) = x^\mu + p^\mu \tau + i \sum_{n\neq 0} \frac{1}{n} \alpha_n^\mu \, e^{-in\tau} \, Cos(n\sigma) \tag{52}$$

$$or \quad \partial_\pm X^\mu = \frac{1}{2} \sum_{-\infty}^{+\infty} \alpha_n^\mu \, e^{-in(\tau\pm\sigma)} \tag{53}$$

$$and \quad [\,\alpha_m^\mu, \alpha_n^\nu\,] = m\,\delta_{m+n}\,\eta^{\mu\nu} \tag{54}$$

For fermions, we have both Neveu-Schwartz (NS) and Ramond (R) boundary conditions. The former states are the bosonic states and the latter is classified as the fermionic states. In the NS sector, the fermions are quantised as

$$\psi_\pm^{\mu,j}(\sigma,\tau) = \frac{1}{\sqrt{2}} \sum_{r\in Z+\frac{1}{2}} b_r^{\mu,j} \, e^{-ir(\sigma\pm\tau)} \tag{55}$$

$$and \quad \phi_\pm^{\mu,k}(\sigma,\tau) = \frac{1}{\sqrt{2}} \sum_{r\in Z+\frac{1}{2}} b_r^{\prime\mu,k} \, e^{-ir(\sigma\pm\tau)}. \tag{56}$$

The creation phase ambiguity is removed by noting $b'_{-r} = -b_r^{\prime\dagger}$. Anticommutators, which do not vanish, are

$$\left\{ b_r^{\mu,j}, b_s^{\nu,j'} \right\} = \delta_{r+s}\,\eta^{\mu\nu}\delta_{jj'} \tag{57}$$

$$and \quad \left\{ b_r^{\prime\mu,k}, b_s^{\prime\nu,k'} \right\} = -\delta_{r+s}\,\eta^{\mu\nu}\delta_{kk'}. \tag{58}$$

In the Ramond sector, the mode expansions are,

$$\psi_\pm^{\mu,j}(\sigma,\tau) = \frac{1}{\sqrt{2}} \sum_{-\infty}^{\infty} d_m^{\mu,j} \, e^{-im(\sigma\pm\tau)}, \tag{59}$$

$$and \quad \phi_\pm^{\mu,k}(\sigma,\tau) = \frac{1}{\sqrt{2}} \sum_{-\infty}^{\infty} d_m^{\prime\mu,k} \, e^{-im(\sigma\pm\tau)}. \tag{60}$$

There is a similar ansatz for the creation operators, $d'_{-m} = -d_m^{\prime\dagger}$. Anticommutators are,

$$\left\{ d_m^{\mu,j}, d_n^{\nu,j'} \right\} = \delta_{m+n}\,\eta^{\mu\nu}\delta_{jj'}, \tag{61}$$

$$and \quad \left\{ d_n^{\prime\mu,k}, d_m^{\prime\nu,k'} \right\} = -\delta_{n+m}\,\eta^{\mu\nu}\delta_{kk'}. \tag{62}$$

Using notations and procedure laid down in the text book [17], we calculate the Virasoro generators

$$L_m = \frac{1}{\pi}\int_{-\pi}^{\pi} d\sigma e^{im\sigma} T_{++} \tag{63}$$

$$= \frac{1}{2}\sum_{-\infty}^{\infty} : \alpha_{-n}\cdot\alpha_{m+n} :$$

$$+ \frac{1}{2}\sum_{r\in z+\frac{1}{2}} (r+\frac{1}{2}m) : (b_{-r}\cdot b_{m+r} - b'_{-r}\cdot b'_{m+r}) : \quad NS \tag{64}$$

$$= \frac{1}{2}\sum_{-\infty}^{\infty} : \alpha_{-n}\cdot\alpha_{m+n} :$$

$$+ \frac{1}{2}\sum_{n=-\infty}^{\infty} (n+\frac{1}{2}m) : (d_{-n}\cdot d_{m+n} - d'_{-n}\cdot d'_{m+n}), : \quad R \tag{65}$$

$$G_r = \frac{\sqrt{2}}{\pi}\int_{-\pi}^{\pi} d\sigma e^{ir\sigma} J_+ = \sum_{n=-\infty}^{\infty} \alpha_{-n}\cdot\left(e^j b_{n+r,j} - e^k b'_{n+r,k}\right), \qquad NS \tag{66}$$

$$F_m = \sum_{-\infty}^{\infty} \alpha_{-n}\cdot\left(e^j b_{n+r,j} - e^k b'_{n+r,k}\right). \quad R \tag{67}$$

These generators satisfy the Super Virasoro algebra with central charge C=26,

$$[L_m, L_n] = (m-n)L_{m+n} + \frac{C}{12}(m^3-m)\delta_{m,-n}, \tag{68}$$

$$[L_m, G_r] = (\frac{1}{2}m - r)G_{m+r}, \qquad NS \tag{69}$$

$$\{G_r, G_s\} = 2L_{s+r} + \frac{C}{3}(r^2-\frac{1}{4})\delta_{r,-s}, \tag{70}$$

$$[L_m, F_n] = (\frac{1}{2}m - n)F_{m+n}, \qquad R \tag{71}$$

$$\{F_m, F_n\} = 2L_{m+n} + \frac{C}{3}(m^2-1)\delta_{m,-n}, \quad m \neq 0. \tag{72}$$

The reappearance of the central charge 26, as for an open bosonic string is interesting and need further elaboration.

The central charge can be calculated from the leading divergence of the vacuum expectation value of the product of the two energy momentum tensors at two world sheet points z and ω. Using

$$< X^\mu(z), X^\nu(\omega) >\sim \eta^{\mu\nu}\, log(z-\omega), \quad < \psi^{\mu,j}(x), \psi^{\nu,j'}(\omega) >\sim \eta^{\mu\nu}\delta^{jj'}(z-\omega)^{-1} \tag{73}$$

and

$$< \phi^{\mu,j}(z), \phi^{\nu}_{k'}(\omega) >\sim \eta^{\mu\nu}\delta^k_{k'}(z-\omega)^{-1}, \tag{74}$$

one deduces that

$$2 < T_+(z), T_+(\omega) >\sim C(z-\omega)^{-4} + ... \tag{75}$$

where

$$C = \eta^{\mu}_{\mu} + \frac{1}{2}\eta^{\mu}_{\mu}\delta^{j}_{j} + \frac{1}{2}\eta^{\mu}_{\mu}\delta^{k}_{k} = 26 \tag{76}$$

But a normal superstring should have a central charge 15. To explain this, we observe that the light cone gauge is ghost free. Dropping the helicity suffixes, the light cone vectors $\psi^{\pm}_{j} = \frac{1}{\sqrt{2}}(\psi^{0}_{j} \pm \psi^{3}_{j})$ and $\psi^{\pm}_{j} = \frac{1}{\sqrt{2}}(\psi^{0}_{j} \pm \psi^{3}_{j})$ have their anticommutators with a negative sign. They are the ghost modes which are the extras. The total ghost energy-momentum tensor comes from the (0,3) coordinates.

$$T^{gh}_{l.c}(z) = \frac{i}{2}(\psi^{0j}\partial_Z\psi_{0j} + \psi^{3j}\partial_Z\psi_{3j}) - \frac{i}{2}(\phi^{0k}\partial_Z\phi_{0k} + \phi^{3k}\partial_Z\phi_{3k}) \tag{77}$$

The vacuum correlation function is calculated to be

$$2 < T^{gh}_{l.c}(z),\, T^{gh}_{l.c}(\omega) > \simeq 11(z-\omega)^{-4} \tag{78}$$

The addition of 11 makes the total of 26. Without them, the central charge is 15 as usual. These terms in equation (15) behaves like the superconformal ghost action and should not show up in the construction of physical states.

6 Bosonic and Fermionic Physical States

A physical bosonic state $|\Phi>$ can be conveniently constructed by operating the Virasoro Generators L_m and G_r. They satisfy

$$\begin{aligned} (L_o - 1)|\Phi> &= 0, && (79)\\ L_m|\Phi> &= 0, \quad for \quad m > 0 && (80)\\ G_r|\phi> &= 0. \quad for \quad r > 0. && (81) \end{aligned}$$

The conditions should make the string free of ghosts. It can be seen in a simple way. Applying L_0 condition , the state $\alpha^{\mu}_{-1}|0,k>$ is massless and the L_1 constraint gives the Lorentz condition $k^{\mu}|0,k>$ =0 implying that the photon is transverse and the Gupta-Bleuler [21] impose $\alpha^{0}_{-1}|\Phi>$ =0. Applying, in succession, $L_1, L_2, ..$ constraints, one obtains $\alpha^{0}_{m}|\Phi>$=0. Further, since $\left[\alpha^{0}_{-1}, G_{r+1}\right]|\Phi>$=0, we should have $b^{0}_{r,j}|\Phi>$=0 and $b'^{0}_{r,k}|\Phi>$=0 as well. Thus there is no contributions from the time components.

Even though no negative norm state should show, we make it doubly sure, by constructing the zero norm states or the 'null' physical states. Due to G.S.O. condition, given in a latter section, the physical states are obtained by operating product of even number of G_r's. There exists generic states $|\tilde{\chi}>$ such that $L_0|\tilde{\chi}_n>= (1-n)|\tilde{\chi}_n>$. n=0 gives tachyon, $|\tilde{\chi}_0>$. The next higher state $|\tilde{\chi}_{\frac{1}{2}}>$ is another tachyon and is projected out by G.S.O.. So the next allowed state is

$$|\Phi>= L_1|\tilde{\chi}_1> + \lambda G_{-\frac{1}{2}}G_{-\frac{1}{2}}|\tilde{\chi}_1> . \tag{82}$$

But $G_{-\frac{1}{2}}G_{-\frac{1}{2}} = L_1$, so in general, the state is

$$|\Phi>= L_1|\tilde{\chi}_1>, \tag{83}$$

which satisfies $(L_0 - 1)|\Phi >=0$ and is true if $L_0|\tilde{\chi}_1 >=0$. The norm $< \Phi|\Phi >=< \tilde{\chi}_1|L_1 L_{-1}|\tilde{\chi}_1 >= 2. < \tilde{\chi}_1|L_0|\tilde{\chi}_1 >=0$. This is a null state. Arguing as above, the next excited state is simplified to be

$$|\Phi >= (L_2 + rL_1^2)|\tilde{\chi}_2 > \tag{84}$$

The condition $(L_0 - 1)|\phi >=0$ is satisfied if $(L_0 + 1)|\tilde{\chi}_2 >=0$. The physical state condition $L_1|\Phi >=0$ gives the value $\gamma = \frac{3}{2}$. The norm, using Virasoro algebra anomaly term, is easily obtained as

$$< \Phi|\Phi >= \frac{1}{2}(C - 26) \tag{85}$$

This is negative for C is less than 26 and 'null' if C=26 which is the central charge. It is easy to check that $L_2|\phi >= 0$ for C=26.

In the Ramond fermionic sector, the arguments run very much the same. The physical state $|\Psi >$ conditions are

$$(L_0 - 1)|\Psi >_\alpha = (F_0 - 1)(F_0 + 1)|\Psi_\alpha >= 0 \tag{86}$$

and

$$L_m|\Psi >= F_m|\Psi >= 0 \qquad for\ m > 0 \tag{87}$$

The first excited state $|\Psi >= L_{-1}|\tilde{\chi}_1 >$ is of zero norm. The next excited state

$$|\Psi >= (L_2 + \frac{3}{2}L_{-1}^2|\tilde{\chi}_2 > \tag{88}$$

has the norm $\frac{1}{2}(C - 26)$ and becomes 'null' for C=26. Thus there is no negative norm states in the Ramond sector as well. We shall now discuss about the vacuum state tachyons. There is one in bosonic sector and one pair among Ramond fermions. The tachyonic energy of the former is $< 0|(L_0 - 1)^{-1}|0 >_{NS}$. This is completely cancelled out by $- < 0|(F_0 + 1)^{-1}(F_0 - 1)^{-1}|0 >_R$, the negative sign is due to the fermionic loop. This is what happens in supersymmetric cancellation of divergences.

7 Conformal and Superconformal Ghosts and the Nilpotent BSRT Charge

The anomalies of Virasoro algebra, namely the terms containing the central charge C, cancel out if the Faddeev-Popov ghost action (FP) [22] is added to the action of equation (5)

$$S_{FP} = \frac{1}{\pi}\int \left(c^+\partial_- b_{++} + c^-\partial_+ b_{--}\right) d^2\sigma \tag{89}$$

b and c are the ghost fields. They satisfy anticommutation relation and are quantised with the mode expansion,

$$c^\pm = \sum_{n=-\infty}^{\infty} c_n\, e^{-in(\tau\pm\sigma)} \tag{90}$$

$$b_{\pm\pm} = \sum_{n=-\infty}^{\infty} b_n\, e^{-in(\tau\pm\sigma)}, \tag{91}$$

by

$$\{c_m, b_n\} = \delta_{m+n}, \ \ \{b_m, b_n\} = \{c_m, c_n\} = 0. \tag{92}$$

As given in reference[15], the energy-momentum Virasoro generators are

$$L_m^{F.P} = \sum_{n=-\infty}^{\infty} (m-n)\, b_{m+n} c_{-n} - a\delta_{m+n}. \tag{93}$$

satisfying the algebra

$$\begin{aligned} [L_m^{FP}, L_n^{FP}] &= (m-n)L_{m+n}^{FP} + \frac{1}{6}(m - 13m^3) + 2am, \\ &= (m-n)L_{m+n}^{FP} - \frac{26}{12}(m^3 - m), \quad when \quad a = 1. \end{aligned} \tag{94}$$

Using Jacobi Identity, one can deduce

$$\left[G_r^{FP}, G_s^{FP}\right] = 2L_{r+s}^{FP} - \frac{26}{3}(r^2 - \frac{1}{14}) \tag{95}$$

Thus the total action of equation (5) added to the F.P. action of equation(89) has the anomalies cancelled in the Virasoro generator and in the theory. The BRST charge which justifies the energy momentum generators (i.e. L_m) constraints on physical states is

$$(Q_1) = \sum_{-\infty}^{\infty}(L_{-m}c_m) - \frac{1}{2}\sum_{-\infty}^{\infty}(m-n) : c_{-m}c_{-n}b_{m+n} : \ - c_0. \tag{96}$$

and independently Q_1^2=0 by construction using Lie algebra [15]. It appears that the superconformal ghosts are needed. But then only the BRST charge will ensure the current generator constraints on the physical states. The total cancellation above is due to the fact that the action (5) has within itself a part which is the superconformal ghost action. This is the light cone part of the action. It is simpler to write it from the action(18) with $\Psi^{\pm} = \frac{1}{\sqrt{2}}(\Psi^0 \pm \Psi^3)$. The action is

$$S_F^{l.c} = -\frac{i}{\pi}\int d^2\sigma \ \bar{\Psi}^+ \rho^\alpha \partial_\alpha \Psi^-. \tag{97}$$

We supplement this Hilbert space by another space where the fermi particles satisfy bose statistics. This is done here by letting $\bar{\Psi}^+$ and Ψ^- to be the function of Grassman variable $\bar{\theta}$ and θ as well. Then we write the action

$$S_F^{l.c} = -\frac{i}{\pi}\int d^2\sigma \int d\bar{\theta} \int d\theta \ \bar{\Psi}^+(\bar{\theta})\ \rho^\alpha \partial_\alpha \Psi^-(\theta). \tag{98}$$

Expanding the fields as

$$\bar{\Psi}^+(\bar{\theta}) = \cdots + \bar{\theta}\bar{\gamma} + \cdots \tag{99}$$

and

$$+2\, i\ \rho^\alpha \Psi^-(\theta) = \cdots + \theta\beta^\alpha + \cdots \tag{100}$$

The Superconformal ghost action is pulled out as

$$S_F^{l.c} = -\frac{1}{2\pi}\int d^2\sigma \ \bar{\gamma}\partial_\alpha\beta_\alpha. \tag{101}$$

The energy momentum tensor is

$$T_{++} = -\frac{1}{4}\gamma\, \partial_+\beta - \frac{3}{4}\beta\partial_+\gamma, \tag{102}$$

and the wave equations are $\partial\gamma = \partial\beta = 0.$

With 'p' integral for Ramond and half integral for NS, the field quanta $\gamma_p,\ \beta_p$ satisfy the only nonvanishing commutator relation

$$[\gamma_p, \beta_r] = \delta_{p+r}. \tag{103}$$

The $L^{l.c.}$-generator is transformed into

$$L_m^{l.c.} = \sum_n (n + \frac{1}{2}m) \ :\beta_{m-n}\gamma_n : \tag{104}$$

equaling that for the superconformal ghosts. All that we need to prove that the nilpotency of BSRT charge is that the conformal dimension of γ is '-1/2' and that of β is '3/2' respectively as can be deduced from equation(104) using equation (103).

The BRST charge which ensures the validity of the current generator constraints on physical states is

$$Q' = \sum_r G_r\gamma_{-r} - \sum_{r,s} \gamma_{-s}\gamma_{-r}\, b_{s+r}. \qquad NS \tag{105}$$

$$Q' = \sum_m F_m\gamma_{-m} - \sum_{m,n} \gamma_{-m}\gamma_{-n}\, b_{m+n}. \qquad R \tag{106}$$

The total BRST charge is

$$Q_{BRST} = Q_1 + Q' \tag{107}$$

Using Virasoro algebra and the field equation $\partial\gamma = \partial\beta = 0$, it is found, after a lot of simplification, that

$$Q^2_{BRST} = 0 \tag{108}$$

The nilpotency of the BSRT charge proves that the theory is free of anomalies and ghosts and that the superstring theory is also unitary.

8 The Mass Spectrum, G.S.O. Operator and Modular Invariance

The mass shell condition is read off from L_0

$$\alpha' M^2 = N^B + N^F_{NS} - 1, \qquad NS, \tag{109}$$

$$= N^B + N^F_{NS} - 1, \qquad R. \tag{110}$$

Here

$$N^B = \sum_{m=1}^{\infty} \alpha_{-m}\, \alpha_m, \tag{111}$$

$$N^F_{NS} = \sum_{r=1/2}^{\infty} r\,(b_{-r} \cdot b_r + b'_{-r} \cdot b'_r), \qquad NS, \tag{112}$$

$$\text{and} \quad N^F_R = \sum_{m=1}^{\infty} m(d_{-m} \cdot d_m + d'_{-m} \cdot d'_m), \qquad R. \tag{113}$$

Due to the coexistence of Ramond and Neveu-Schwartz sectors with periodic and antiperiodic boundary conditions, there exists a G.S.O. operator G [20] for projecting out the unphysical states from the mass spectrum of the NSR model

$$G = \frac{1}{2}(1 + (-1)^{F+F'}), \tag{114}$$

where $F = \sum b_{-r} \cdot b_r$ and $F' = -\sum b'_{-r} \cdot b'_r$. This will eliminate the half integral values from the mass spectrum by choosing G=1, including the tachyon at $\alpha M^2 = -\frac{1}{2}$ in the NS sector.

Assigning the GSO, we proceed to calculate the partition function of the four bosons and forty-four fermions as given in the general action (4). In covariant formulation, the effective number of physical modes is the total number of modes subtracted by the number of invariant constraints. There are two transverse bosonic and two 'bosonic' ghost modes. With four current and energy momentum constraint, the number of effective fermions for partitioning is forty.

It is easy to construct the modular invariant partition function for the two physical bosons, namely

$$\mathcal{P}_B(\tau) = (Im\,\tau)^{-2}\eta^{-2}(\tau)\bar{\eta}^{-2}(\tau). \tag{115}$$

$\eta(\tau)$ is the Dedekind eta function and the definition $q = e^{i\pi\tau}$ will be used later. The partion function for the two bosonic ghosts can be calculated from the equations for the Hamiltonian [22].

$$\mathcal{P}_{\text{ghost}} = \mid \frac{\Theta_2(\tau)}{\eta(\tau)} \mid^4 . \tag{116}$$

Θ's are the Jacobi theta functions. $\eta(\tau)2\pi = \Theta_1^{'\frac{1}{3}}$ For the forty fermions, we let $b^i_{r,k}$ be the annihilation fermion quanta. 'i' running from 1 to 8 and the k, a $SO(5)$ index running from 1 to 5. The $SO(6) \times SO(5)$ invariant (Note that SO(6) has $2^3 = 8$ space time fermionic modes) Hamiltonian in the NS sector is $H^{NS} = \sum_{k=1}^{5} H^{NS}_k$, with $H^{NS}_k = \sum_r r b^{\dagger i}_{rk} b^i_{rk} - 8/48$. We see that $[H_k, H_{k'}] = 0$. So there are five replicas of the group of eight. The partition function of the forty fermions will be the partition function of the eight fermions raised to the power of five.

The path integral functions of Seiberg and Witten [23] for the eight fermions, obtained from the Kaku's formula [24] for a single fermion is

$$A((--), \tau) = (\Theta_3(\tau)/\eta(\tau))^4\,, \tag{117}$$

This is normalised to one. The other three spin structures are related to the above function,

$$A((+-),\tau) = A((--),\frac{\tau}{1+\tau}) = -(\Theta_2(\tau)/\eta(\tau))^4, \tag{118}$$

$$A((-+),\tau) = A((+-),-\frac{1}{\tau}) = -(\Theta_4(\tau)/\eta(\tau))^4\,, \tag{119}$$

and

$$A((++).\tau) = 0. \tag{120}$$

The sum of all spin structures is, therefore,

$$A(\tau) = (\Theta_3(\tau)/\eta(\tau))^4 - (\Theta_2(\tau)/\eta(\tau))^4 - (\Theta_4(\tau)/\eta(\tau))^4. \tag{121}$$

Consider the amplitude

$$A_N(\tau) = \eta^4(\tau)A(\tau) = \Theta_3^4(\tau) - \Theta_2^4(\tau) - \Theta_4^4(\tau). \tag{122}$$

Since

$$\Theta_3(\tau+1) = \Theta_4(\tau), \qquad \Theta_2(\tau+1) = e^{i\pi/4}\Theta_2(\tau), \qquad \Theta_4(\tau+1) = \Theta_3(\tau), \tag{123}$$

we have the desired relation

$$A_N(\tau+1) = -A_N(\tau). \tag{124}$$

The invariance $\tau \to -1/\tau$ is built in, in the ratios of the theta functions of each structure.

The modular invariant partition function for all the forty fermions is

$$\mathcal{P}_F(\tau) =\mid A(\tau) \mid^{10}, \tag{125}$$

and total partition function for the model is the integral of the product,

$$Z = \int \mathcal{P}_B \mathcal{P}_{ghost} \mathcal{P}_F d\tau. \tag{126}$$

We notice that due to the famous Jacobi relation, the $A(\tau)$ equals zero, so the modular invariant partition function Z vanishes. Thus this necessary condition for space-time supersymmetry is satisfied. At the same time, the constant in the power to which 'q' is raised, is still, $-\frac{2}{24} - \frac{2}{24} - \frac{40}{48} = -1$, confirming the correctness of our calculation above.

9 Supersymmetric Standard Model

As such, the theory would have been worthless, had it not been able to make contact with low energy phenomenology. The most successful theory of elementary particle physics, has been the supersymmetric Standard Model. Besides explaining almost all available and established data, it has successfully predicted a grand unification of the three gauge couplings of the $SU_C(3) \otimes SU_L(2) \otimes U_Y(1)$ at a scale of about 2×10^{16} GeV.

We have an action, equation (5), which is invariant under the interesting(world sheet) symmetry group $SO(6) \otimes SO(5)$. What about the gauge symmetry? For this purpose, one has to examine the zero mass vector bosons. Following Li [25], we seek adjoint representation tensors, which are antisymmetric in both Lorentz (μ, ν) and O(n) indices (i,j). For simplicity, let us denote the creation quanta of ψ_i^μ as $b_i^{\mu\dagger}$. Then $[L_0,\ b_i^{\mu\dagger}] = -\frac{1}{2}b_i^{\mu\dagger}$. The desired field strength tensors are [25]

$$F_{\mu\nu,ij} = \epsilon_{\mu\nu\lambda\sigma}\left(b_i^{\lambda\dagger}b_j^{\sigma\dagger} - b_j^{\lambda\dagger}b_i^{\sigma\dagger}\right)|0>, \tag{127}$$

with $\frac{1}{2}n(n-1)$ gauge bosons W_{ij}^μ, given by

$$F_{ij}^{\mu\nu} = \partial^\nu W_{ij}^\mu - \partial^\mu W_{ij}^\nu + g(W_{ik}^\nu W_{kj}^\mu - W_{ik}^\nu W_{kj}^\mu). \tag{128}$$

It is clear that $L_o F_{ij}^{\mu\nu}$ =0 and gauge bosons are massless. When (i,j) run from 1 to 6, as in ψ_i^μ ,we get the adjoint $\underline{15}$ of the gauge group SO(6). Similarly for ϕ_k^μ, where k runs from 7 to 11, the field strength tensor is like $F_{kl}^{\mu\nu}$ which is $\underline{10}$ of SO(5) of the gauge bosons. So from the massless spectrum, we find that the gauge symmetry of the string is $SO(6) \otimes SO(5)$, which is the same as that of action.

We have to descend to the standard model group $SU_C(3) \otimes SU_L(2) \otimes U_Y(1)$ from $SO(6) \otimes SO(5)$, it can be normally done by introducing Higgs which breaks gauge symmetry and supersymmetry. However, one uses the method of symmetry breaking by using Wilson's lines, supersymmetry remains in tact but the gauge symmetry is broken. This Wilson loop is

$$U_\gamma = P\ exp(\oint_\gamma A_\mu\ dx^\mu) \tag{129}$$

P represents the ordering of each term with respect to the closed path γ. SO(6)=SU(4) descends to $SU_C(3) \otimes U_{B-L}(1)$. This breaking can be accomplished by choosing one element of U_o of SU(4), such that $U_o^2 = 1$. These elements generate the permutation group Z_2. Thus, without breaking supersymmetry,

$$\frac{SO(6)}{Z_2} = SU_C(3) \otimes U_{B-L}(1). \tag{130}$$

Similarly

$$\frac{SO(5)}{Z_2} = SU(2)_L \otimes U_R(1). \tag{131}$$

Combined operation will lead to

$$\frac{SO(6) \otimes SO(5)}{Z_2 \otimes Z_2} = SU_C(3) \otimes SU(2)_L \otimes U_{B-L}(1) \otimes U_R(1). \tag{132}$$

Such a model is usually used in discussing low energy phenomena. But this is not the standard model. We have an additional U(1). We recall that in E_6 symmetry breaking, there is reduction of rank by one. So we hope to achieve, likewise,

$$\frac{SO(6) \otimes SO(5)}{Z_3} = SU_C(3) \otimes SU(2)_L \otimes U_Y(1). \tag{133}$$

We describe Z_3 by

$$g(\theta_1,\theta_2,\theta_3) = (\frac{2\pi}{3} - 2\theta_1, \frac{2\pi}{3} + \theta_2, \frac{2\pi}{3} + \theta_3). \tag{134}$$

For the first Wilson loop, the angle integral for θ_1 is from $\frac{2\pi}{9}$ to $\frac{2\pi}{3} - \frac{4\pi}{9} = \frac{2\pi}{9}$, so that the loop integral vanishes. For θ_2, the second loop is described by the length parameter R and this angle varies from 0 to $2\pi - \frac{2\pi}{3} = \frac{4\pi}{3}$. The angle θ_3 for the remaining loop varies from 0 to $2\pi - \frac{2\pi}{3} = \frac{4\pi}{3}$ with the same R. We take the polar components of the gauge fields as nonzero constants like

$$gA^{15}_{\theta_2} = \vartheta_{15} \quad for \quad SO(6) = SU(4), \tag{135}$$

for which the diagonal generator t_{15} breaks the symmetry and

$$g' A^{10}_{\theta_3} = \vartheta'_{10}. \tag{136}$$

for the SO(5), the diagonal generator being t'_{10}. The generators of both the SO(6) and the SO(5) are 4×4 matrices. We can write the Z_3 group as

$$T = T_{\theta_1} T_{\theta_2} T_{\theta_3}. \tag{137}$$

We have T_{θ_1} =1 and this leaves the symmetry $SU(3) \otimes SU(2)$ untouched by choice. The next

$$T_{\theta_2} = exp\left(i\, t_{15} \int_o^{\frac{4\pi}{3}} \vartheta_{15} R d\theta_2 \right), \tag{138}$$

and breaks the SU(4) symmetry if $T_{\theta_2} \neq 1$. Similarly

$$T_{\theta_3} = exp\left(i\, t'_{10} \int_o^{\frac{4\pi}{3}} \vartheta'_{10} R d\theta_3 \right), \tag{139}$$

breaks the SO(5) symmetry if $T_{\theta_3} \neq 1$.

But in Z_3, we have the product,

$$T_{\theta_2} T_{\theta_3} = exp\left(i \int_o^{\frac{4\pi}{3}} (\vartheta'_{10} t'_{10} + \vartheta_{15} t_{15}) R d\theta \right). \tag{140}$$

Since ϑ_{15} and ϑ'_{10} are arbitrary constants, we can choose in such a way that $t_{15}\vartheta_{15} + t'_{10}\vartheta'_{10} = 0, \frac{3}{2R}$........ Thus T=1, and the standard model gauge group (133) is obtained. The rank has been reduced by one. Since Z_3 has three generators, the Euler number is 3. There are three generations in $SO(6) \otimes SO(5) = Z_3 \otimes SU(3)_C \otimes SU_L(2) \otimes U_Y(1)$ gauge symmetry. This is a very important and unique result.

10 Supersymmetric Standard Model Particle Spectrum

Let us now discuss in detail the massless particle spectrum of one generation. The B_μ, the hypercharge vector, is simply given by the state

$$|\phi(p)>= \alpha^\mu_{-1}|0,p> B_\mu. \tag{141}$$

L_0 condition tells that $B_\mu(p)$ is massless and the L_1 condition gives the Lorentz condition $p^\mu\, B_\mu = 0$.

In NS sector, the product of the objects like $b^\mu_{-\frac{1}{2},i}\, b^\mu_{-\frac{1}{2},j}$ are also massless. Dropping the suffix $-\frac{1}{2}$, we construct tensors (i,j = 1,2,3), expressible in terms of the color vector fields A^μ_{ij},

$$\begin{aligned} G^{\mu\nu}_{ij} &= \left(b^\mu_i b^\nu_j - b^\nu_i b^\mu_j - \frac{2}{3}\delta_{ij} b^\mu_l b^\nu_l\right) && (142)\\ &= \partial^\nu A^\mu_{ij} - \partial^\mu A^\nu_{ij} + g_3\left(A^\mu_{il}A^\nu_{jl} - A^\nu_{il}A^\mu_{jl}\right). && (143)\end{aligned}$$

The eight gluons are obtained by using the eight Gell-Mann λ_l matrices.

$$V^\mu_l = (\lambda_l)_{ij} A^\mu_{ij}. \tag{144}$$

Similarly for (i,j = 7,8), we get the W^μ-mesons

$$\begin{aligned} W^{\mu\nu}_{ij} &= \left(b'^\mu_i b'^\nu_j - b'^\nu_i b'^\mu_j - \delta_{ij} b'^\mu_i b'^\nu_j\right) && (145)\\ &= \partial^\nu W^\mu_{ij} - \partial^\mu W^\nu_{ij} + \left(W^\mu_{il}W^\nu_{jl} - W^\nu_{il}W^\mu_{jl}\right), && (146)\end{aligned}$$

with

$$W^\mu_l = (\tau_l)_{ij} W^\mu_{ij}, \tag{147}$$

where τ_l, l=1,2,3 are the 2×2 isospin matrices. Of the remaining b^μ_j, j=4,5,6 and b'^μ_k, k=9,10,11, we can form six Lorentz scalars $\epsilon_{ijl} b^\lambda_j b^\lambda_l$ and $\epsilon_{ikm} b'^\lambda_k b'^\lambda_m$. Let us call them $\phi_1, ..., \phi_6$. The Higgs bosons are

$$\phi_H = \begin{pmatrix} \phi_1 + i\phi_2 \\ \phi_3 + i\phi_4 \end{pmatrix}, \tag{148}$$

leaving behind a scalar

$$\varphi_s = \phi_5 + i\phi_6. \tag{149}$$

Let us consider the fermion sector. With Γ^μ as the usual 4x4 Dirac matrices and u(p) being the fermion wave function, the states $\Gamma^\mu d_{-1,j,\mu}|0,p>u(p)$ and $\Gamma^\mu d'_{-1,k,\mu}|0,p>$

$u(p)$ are all massless states. Broadly j=1,..,6 are the colour sector and k=7,..,11 are the electroweak sector. The current generator condition gives $\Gamma^{\mu} p_{\mu} u(p) = 0$, the Dirac equation for each of them.

We have to find a way to put these fermions into electroweak and colour groups of one generation. Let us drop the Dirac indices, the suffix -1, and prefix primes and supply a factor 2 such that

$$\{d_i, d_j\} = 2\delta_{ij} \tag{150}$$

Since SO(4) can come from either SO(6) or SO(5) [25], we can choose any four of them and define new operators b's as follows

$$b_1 = \frac{1}{2}(d_1 + id_2), \quad b_1^* = \frac{1}{2}(d_1 - id_2), \quad b_2 = \frac{1}{2}(d_3 + id_4), \quad and \quad b_2^* = \frac{1}{2}(d_3 - id_4). \tag{151}$$

They satisfy the algebra (i,j=1,2),

$$\{b_i, b_j\} = 0, \quad \{b_i^*, b_j^*\} = 0 \ \ and \ \ \{b_i, b_j^*\} = 2\delta_{ij} \tag{152}$$

This is the U(2) group, identified here as the isospin group. We now construct a table giving the assignment of all fermions (Table 1).

Table 1: Particle Spectra, States and Quantum Numbers (SM)

Particles	States	I_3	Y	Q
e_R	$\Gamma \cdot d'_7 \vert 0>_\alpha$	0	-2	-1
ν_L	$\frac{1}{\sqrt{2}}\Gamma \cdot (d'_8 + id'_9)\vert 0>_\alpha$	$1/2$	-1	0
e_L	$\frac{1}{\sqrt{2}}\Gamma \cdot (d'_{10} + id'_{11})\vert 0>_\alpha$	$-1/2$	-1	-1
$u_R^{a,b,c}$	$\Gamma \cdot d_{1,2,3}\vert 0>_\alpha$	0	4/3	2/3
$d_R^{a,b,c}$	$\Gamma \cdot d_{4,5,6}\vert 0>_\alpha$	0	-2/3	-1/3
$u_L^{a,b,c}$	$\frac{1}{\sqrt{2}}\Gamma \cdot (d_1 + id_4)\vert 0>_\alpha$, $\frac{1}{\sqrt{2}}\Gamma \cdot (d_2 + id_5)\vert 0>_\alpha$, $\frac{1}{\sqrt{2}}\Gamma \cdot (d_3 + id_6)\vert 0>_\alpha$,	$1/2$	$1/3$	$2/3$
$d_L^{a,b,c}$	$\frac{1}{\sqrt{2}}\Gamma \cdot (d_1^c + id_4^c)\vert 0>_\alpha$, $\frac{1}{\sqrt{2}}\Gamma \cdot (d_2^c + id_5^c)\vert 0>_\alpha$, $\frac{1}{\sqrt{2}}\Gamma \cdot (d_3^c + id_6^c)\vert 0>_\alpha$	$-1/2$	$1/3$	$-1/3$

Once the weak isospins are fixed, the hypercharge can be assigned or calculated from the charge also. There are fifteen fermions with 30 degrees of freedom matching the 30 bosonic degrees of freedom from above equations. We now proceed to construct the Hamiltonian which is supersymmetric and $SU_C(3) \otimes SU_L(2) \otimes U_Y(1)$ invariant following the ideas presented by one of us [14]. The vector fields denoted by V_μ^l, l=1,..,8, are gluon fields. l=9 is the $U_Y(1)$ field and l=10,11,12 stand for W-mesons fields. We shall use the temporal gauge [26], where V_0^l =0. For each V_i^l, there are electric field strength E_i^l and magnetic field strength B_i^l. Following Nambu [27], the combination

$$F_i^l = \frac{1}{\sqrt{2}}\left[E_i^l + B_i^l\right], \tag{153}$$

satisfies the only nonvanishing equal time commutation relation

$$\left[F_i^{\dagger l}(x), F_j^m(y)\right] = i\delta^{lm}\epsilon_{ijk}\partial^k(x-y). \tag{154}$$

We then construct Wilson's loop line integrals to convert the ordinary derivatives acting on fermion fields to respective gauge covariant derivatives. The phase function for the colour is

$$U_C(x) = exp\left(ig\int_0^x \sum_{l=1}^{\infty}\lambda^l V_i^l dy_i\right). \tag{155}$$

Denoting

$$Y(x) = g'\int_0^x B_i dy_i, \tag{156}$$

the isospin phase functions are

$$\begin{aligned}
U_Q(x) &= exp\left(\frac{ig}{2}\int_0^x \tau\cdot\mathbf{W}_i dy_i - \frac{i}{6}Y(x)\right), && (157)\\
U(x) &= exp\left(\frac{ig}{2}\int_0^x \tau\cdot\mathbf{W}_i dy_i - \frac{i}{2}Y(x)\right), && (158)\\
U_1(x) &= exp\left(-\frac{2i}{3}Y(x)\right), && (159)\\
U_2(x) &= exp\left(\frac{i}{3}Y(x)\right), && (160)\\
\textit{and}\quad U_R(x) &= exp\left(iY(x)\right) && (161)
\end{aligned}$$

The ψ^l will denote the fermions, l=1,..,6 refer to the coloured quark doublets. The sum of the products

$$\begin{aligned}
\sum_{l=1}^{6} F_i^{l\dagger}\cdot\psi^l &= \sum_{l=1}^{3}\left(F_i^{*l}, F_i^{*l+3}\right)\begin{pmatrix}\psi^l\\ \psi^{l+3}\end{pmatrix} && (162)\\
&= \sum_{l=1}^{3}\left(F_i^{*l}, F_i^{*l+3}\right)U_Q\,U_C\begin{pmatrix}u_L^l\\ d_L^{l+3}\end{pmatrix} && (163)
\end{aligned}$$

The singlet phased colour quarks are

$$\begin{aligned}
\psi^l &= U_1 U_C u_R^l, \quad l = 7,8,9\ and && (164)\\
\psi^l &= U_2 U_C d_R^l, \quad l = 10,11,12. && (165)
\end{aligned}$$

The Q=1 singlet, φ_s and the e_R are phased as

$$\phi_s = U_R\varphi_s, \qquad \psi_R = U_R e_R. \tag{166}$$

The Higgs doublets ϕ and the lepton doublet ψ^l are phased like

$$\phi = U\phi_H \qquad \psi_L = U\begin{pmatrix} \nu_e \\ e^- \end{pmatrix}. \tag{167}$$

In writing down the supersymmetric charge Q, we shall use 2×2 matrices, σ^μ, $\mu = 0, 1, 2, 3$; $\sigma^0 = I$ and σ's are the three Pauli spin matrices,

$$Q = \int d^3x \left[\sum_{l=1}^{12} \left(\sigma \cdot \mathbf{F}^{\dagger l} \psi^l(x) \right) + \sigma^\mu \partial_\mu \phi^\dagger(x) \cdot \psi_L(x) + \sigma^\mu \partial_\mu \phi_s^\dagger(x) \cdot \psi_L(x) + \mathcal{W}\,\psi^1 \right]. \tag{168}$$

where $\mathcal{W} = \sqrt{\lambda}\,[\phi^\dagger\phi - v^2\,]$ is the electroweak symmetry breaking term with $\sigma = \sqrt{2}v = 246\ GeV$. We use the usual commutators and anticommutators at equal times and $\sigma_i\sigma_j = \delta_{ij} + i\epsilon_{ijk}\sigma_k$ and obtain

$$\left\{ Q^\dagger_\alpha, Q_\beta \right\} = (\sigma^\mu P_\mu)_{\alpha\beta}, \tag{169}$$

P_0 is the Hamiltonian H and P is the total momentum. We get the $SU_C(3) \otimes SU_L(2) \otimes U_Y(1)$ invariant Hamiltonian as

$$\begin{aligned} H &= \int d^3x\, [\sum_{l=1}^{12} \frac{1}{2} \left(\mathbf{E}^{l2} + \mathbf{B}^{l2} + i\psi^{l\dagger}(x)\sigma \cdot \nabla\psi^l(x) \right) \\ &+ \pi_H^\dagger(x)\pi_H(x) + (\nabla\phi^\dagger) \cdot \nabla\phi + i\psi_L^\dagger(x)\sigma \cdot \nabla\psi_L(x) + i\psi_R^\dagger(x)\sigma \cdot \nabla\psi_R(x) + \mathcal{W}^2\,] \\ &+ \int d^3x \left[\pi_s^\dagger(x)\pi_s(x) + (\nabla\phi_s^\dagger) \cdot \nabla\phi_s \right] \end{aligned} \tag{170}$$

and similarly lengthy expression for P [28]. The last two terms involving Y=2 meson is absent in the standard model. But the Hamilton's equation of motion for the extra meson, obtained from the total Hamiltonian, is

$$\left(\frac{\partial^2}{\partial t^2} - \nabla^2 \right) \phi_s = 0. \tag{171}$$

and the unwanted objects decouples from other standard model particles. Supersymmetric transformations exhibiting the superpartners are

$$\delta\psi^l = \sigma \cdot \mathbf{F}^l\,\epsilon + \delta_{l1}\mathcal{W}\,\epsilon, \tag{172}$$

$$\delta\mathbf{F}_i^l = -i\,\bar{\epsilon}\,(\sigma \times \nabla)_i\,\psi^l, \tag{173}$$

$$\delta\psi_L^l = \sigma^\mu\partial_\mu\phi\,\epsilon, \tag{174}$$

$$\delta\phi = -i\bar{\epsilon}\psi_L, \tag{175}$$

$$\delta\psi_R = \sigma^\mu\partial_\mu\phi_s\,\epsilon, \tag{176}$$

$$\delta\phi_s = -\,i\,\bar{\epsilon}\,\psi_L. \tag{177}$$

The only particle, not in the standard model, ϕ_s, is noninteracting and decoupled.

In summary of this section, we wish to emphasise that the proposed superstring excitations of zero mass fall into the standard model particles in a very natural way. The bosons are themselves, the superpartners of the fermions and vice versa. It is not surprising that the exotic SUSYs are not observed in the electroweak scale and the present formulation provides a justification for their absence.

11 Conclusion

We have established a new theory and have been able to obtain very important results of Supersymmetric Standard Model. All our result follow from the action of equation (5). It is natural to reflect on the roll of the Nambu-Goto four dimensional superstring action, equation (18) will play in formulation of some complete physical theory.

Let us quantise the superpartner Ψ^μ field. In NS-R formulation,

$$\Psi^\mu_\pm = \frac{1}{\sqrt{2}} \sum_{r \in Z + \frac{1}{2}} B^\mu_r \, e^{-ir(\tau \pm \sigma)}, \quad NS \tag{178}$$

$$= \frac{1}{\sqrt{2}} \sum_m D^\mu_m \, e^{-im(\tau \pm \sigma)}. \quad R \tag{179}$$

The anticommutation relations are

$$\{B^\mu_r,\, B^\nu_s\} = \eta^{\mu\nu}\delta_{r+s}, \qquad \{B^\mu_m,\, B^\nu_n\} = \eta^{\mu\nu}\delta_{m+n}. \tag{180}$$

Some admissible Fock space states are

$$NS \;\; Eigenstates \; : \quad \prod_{n,\mu} \prod_{m,\nu} \{\alpha^\mu_{-n}\} \{B^\nu_{-m}\} \,|0>, \tag{181}$$

$$R \;\; Eigenstates \; : \quad \prod_{n,\mu} \prod_{m,\nu} \{\alpha^\mu_{-n}\} \{D^\nu_{-m}\} \,|0>u. \tag{182}$$

u is a Spinor. GSO projections are necessary for the NS states. The first massless ground state in NS (from the L_0 condition) is a tensor

$$B^\mu_{-\frac{1}{2}} B^\nu_{-\frac{1}{2}} |0, p> . \tag{183}$$

This can be cast into sum of three terms

$$\frac{1}{2}\left(B^\mu_{-\frac{1}{2}} B^\nu_{-\frac{1}{2}} - B^\nu_{-\frac{1}{2}} B^\mu_{-\frac{1}{2}}\right) + \frac{1}{2}\left(B^\mu_{-\frac{1}{2}} B^\nu_{-\frac{1}{2}} + B^\nu_{-\frac{1}{2}} B^\mu_{-\frac{1}{2}} - 2B_{-\frac{1}{2}} \cdot B_{-\frac{1}{2}}\right) + B_{-\frac{1}{2}} \cdot B_{-\frac{1}{2}}. \tag{184}$$

Due to anticommutivity of the creation operators $B_{-\frac{1}{2}}$'s, the 2nd and 3rd terms of equation (184) vanishes, we are left with traceless symmetric tensor

$$h^{\mu\nu}(p) = \frac{1}{2}\left(B^\mu_{-\frac{1}{2}} B^\nu_{-\frac{1}{2}} - B^\nu_{-\frac{1}{2}} B^\mu_{-\frac{1}{2}}\right) |0, p> . \tag{185}$$

The $G_{\frac{1}{2}}|\Phi>= 0$ condition restricts this to have

$$p^\mu h_{\mu\nu} = 0. \tag{186}$$

This is the graviton and is the spin two particle of gravity. The general equation, derived from generalised Einstein action, is $p^\mu h_{\mu\nu} = \frac{1}{2} p^\nu h_{\mu\mu}$. But $h_{\mu\mu}$ is zero in our case which happens to be the special gauge needed for identifying $h_{\mu\nu}$ as the graviton of general relativity.

The Ramond sector zero mass states are very interesting. For a Dirac spinor u_α (α is Dirac index)

$$L_0|0,p> u_\alpha = u_\alpha. \tag{187}$$

But $F_0^2 = L_0$ gives us

$$(F_0+1)(F_0-1)|0,p> u_\alpha = 0. \tag{188}$$

F_0 is essentially $\gamma \cdot p$. So if $F_0|0,p> u_{1\alpha} = u_{1\alpha}$ and $F_0|0,p> u_{2\alpha} = -u_{2\alpha}$, we can relate them as $u_{2\alpha} = \gamma_5 u_{1\alpha}$. We construct spinor which is

$$v_\alpha = (1+\gamma_5)u_{1\alpha} \tag{189}$$

such that

$$F_0|0,p> v_\alpha(p) = 0. \tag{190}$$

We can construct massless spinors, ($L_0 = p_\mu^2$ condition)

$$\begin{aligned} \psi^\mu &= \alpha^\mu_{-1}|0,p> v_\alpha(p) & (191)\\ or\ \ \psi^\mu &= D^\mu_{-1}|0,p> v_\alpha(p). & (192) \end{aligned}$$

From $L_1\psi^\mu = 0$ and $F_1\psi^\mu = 0$ physical state conditions, we have

$$\begin{aligned} p^\mu \psi_\mu &= 0, & (193)\\ \gamma^\mu \psi_\mu &= 0. & (194) \end{aligned}$$

and since $F_0 v_\alpha = 0$, we also have

$$(\gamma \cdot p)\psi_\mu = 0. \tag{195}$$

As shown in the reference [20], the three equations (193) to (195) completely specify a spin $\frac{3}{2}$ particle only without any spin $\frac{1}{2}$ admixture and obey Rarita-Schwinger equation in four dimensions as

$$\epsilon^{\mu\nu\lambda\sigma}\gamma_5\gamma_\rho\partial_\nu\psi_\sigma = 0. \tag{196}$$

Thus there are two bosonic degrees of freedom of graviton matching with the two fermionic degrees of freedom of the super partner 'gravitino'.

We state the supergravity action given in a understandable notations

$$S = \int d^4x\, e \left(-\frac{1}{2\kappa^2} R - \frac{1}{2}\bar{\psi}_\mu \gamma_\nu \gamma^5 D_\sigma \psi_\rho \epsilon^{\mu\nu\sigma\rho} \right) \tag{197}$$

This is the first Supergravity action to be written down.

Finally we speculate that the action (1) explains the low energy phenomenology whereas the Nambu-Goto super action(18) will throw light on the nature of relationship between general relativity, quantum gravity and string theory in four dimension.

References

[1] G. Veneziano, *Nuovo Cim.* **57A**, 190 (1968); *Phys. Rep.* **9C**, 190 (1968).

[2] Y. Nambu, *Lectures at Copenhagen Symposium* (1970); T. Goto, *Prog. Theo. Phys.* **46**, 1566 (1971).

[3] J. L. Gervais and B. Sakita, *Nucl. Phys.*. **B34**, 631 (1971).

[4] C. Lovelace, *Phys. Letts.* **B34**, 500 (1971).

[5] J. Scherck and J. H. Schwartz, *Phys. Letts.* **B81**, 118 (1974).

[6] M. B. Green and J. H. Schwartz, *Phys. Letts.* **B149**, 117 (1974); *ibid* **B151**, 21 (1985).

[7] D. J. Gross. J. A. Harvey, E. Martinec and R. Rohm, *Nucl. Phys.* **B256**, 253 (1986); *ibid* **B267**, 75 (1986).

[8] *String theory in four dimension*, edited by M. Dine, North Holland (1988).

[9] T. Antoniadis, C. Buchas, and C. Kounnas, *Nucl. Phys.* **B289**, 87 (1987); D. Chang and A. Kumar, *Phys. Rev.* **D38**, 1893 (1988); *ibid* **B38**, 3739 (1988).

[10] H. Kawai, D. C. Lewellen, C-H. H. Tye,*Nucl. Phy.***B288**, 1 (1987).

[11] S. James Gates Jr. and W. Siegel,*Phys. Letts.***B206**, 631 (1988); D. A. Depiereux, s. James Gates Jr and Q-Hann Park, *Phys. Letts.* **B224**, 364 (1989); S. Bellucci, D.A. Depiereux and S. James Gates Jr, *Phys. Letts.***B232**, 67 (1989); D. A. Depiereux, S. James Gates Jr and B. Radak, *Phys. letts.* **B236**, 411 (1990).

[12] A. Casper, F. Englert, H. Nicolai and A. Taormina, *Phys. Letts.* **B162**, 121 (1985)

[13] F. Englert, A. Hourt and A. Taormina, *JHEP,* **0108**, 013, 1 (2001).

[14] B. B. Deo, *Phys. Letts.* **B557**, 115 (2003).

[15] B. B. Deo and L. Maharana, *Mod. Phys. Letts.* **A9**, 1939 (2004).

[16] B. B. Deo and L. Maharana, *Gravitation and Cosmology*,**4**, 1 (2003) and hep-th/0212004.

[17] M. B. Green, J. H. Schwartz and E. Witten, *Superstring Theory*, Vol.I, Cambridge University Press, England (1987); M. Kaku, *Introduction to Superstring and M. Theory*, 2nd Edn, Springer, New York (1999).

[18] L. Brink, P. Di Vecchia and P. Howe, *Phys. Lett.* **B65**, 471 (1976); S. Deser and B. Zumino, *Phys. Lett.* **B65**, 369 (1976).

[19] A. Neveu and J. H. Schwartz, *Nucl. Phys.* **B31**, 86 (1971); P. Ramond, *Phys. Rev.* **D3**, 2415 (1971).

[20] F. Gliozzi, J. Scherk and D. Olive, *Phys. Lett.* **B65**, 282 (1976).

[21] S. N. Gupta, *Proc. Phys. Soc.*(London), **A63**, 681 (1950); K. Bleuler, Helva. *Phys. Acta,* **23**, 567 (1950).

[22] L.D. Faddeev and V. N. Popov, *Phys. Lett.* **B25**,29 (1967).

[23] N. Seiberg and E. Witten, *Nucl. Phys.* **B276**, 27 (1986).

[24] M. Kaku,*Strings, Conformal Fields and Topology*, Springer Verlag, New York(1991).

[25] L. F. Li, *Phys. Rev.* **D9**, 1723 (1974).

[26] N. H. Christ and T. D. Lee, *Phys. Rev.* **D22**, 939 (1980).

[27] Y. Nambu, *Supersymmetry and Quasisupersymmetry*, E. F. Reprints (unpublished) 1991.

[28] M. A. Virasoro, *Phys. Rev.* **D1**, 2933 (1970).

INDEX

E

F

G

H

I

L

M

N

O

P

Q

R

S

T

U

V

W